AF361887

Feature Papers in Neglected and Emerging Tropical Disease

Feature Papers in Neglected and Emerging Tropical Disease

Editor

Vyacheslav Yurchenko

Basel • Beijing • Wuhan • Barcelona • Belgrade • Novi Sad • Cluj • Manchester

Editor
Vyacheslav Yurchenko
Life Science Research Centre,
University of Ostrava
Ostrava, Czech Republic

Editorial Office
MDPI
St. Alban-Anlage 66
4052 Basel, Switzerland

This is a reprint of articles from the Special Issue published online in the open access journal *Tropical Medicine and Infectious Disease* (ISSN 2414-6366) (available at: https://www.mdpi.com/journal/tropicalmed/special_issues/NETD_FP).

For citation purposes, cite each article independently as indicated on the article page online and as indicated below:

Lastname, A.A.; Lastname, B.B. Article Title. *Journal Name* **Year**, *Volume Number*, Page Range.

ISBN 978-3-0365-8736-3 (Hbk)
ISBN 978-3-0365-8737-0 (PDF)
doi.org/10.3390/books978-3-0365-8737-0

Contents

Tropical Medicine and Infectious Disease

Article

Kinetoplast Genome of *Leishmania* spp. Is under Strong Purifying Selection

Evgeny S. Gerasimov [1,2,*], Tatiana S. Novozhilova [1], Sara L. Zimmer [3] and Vyacheslav Yurchenko [4,*]

1   Department of Molecular Biology, Lomonosov Moscow State University, 119234 Moscow, Russia
2   Institute for Information Transmission Problems, Russian Academy of Sciences, 127051 Moscow, Russia
3   Department of Biomedical Sciences, University of Minnesota Medical School, Duluth Campus, Duluth, MN 55812, USA
4   Life Science Research Centre, Faculty of Science, University of Ostrava, 710 00 Ostrava, Czech Republic
*   Correspondence: jalgard@gmail.com (E.S.G.); vyacheslav.yurchenko@osu.cz (V.Y.)

Abstract: Instability is an intriguing characteristic of many protist genomes, and trypanosomatids are not an exception in this respect. Some regions of trypanosomatid genomes evolve fast. For instance, the trypanosomatid mitochondrial (kinetoplast) genome consists of fairly conserved maxicircle and minicircle molecules that can, nevertheless, possess high nucleotide substitution rates between closely related strains. Recent experiments have demonstrated that rapid laboratory evolution can result in the non-functionality of multiple genes of kinetoplast genomes due to the accumulation of mutations or loss of critical genomic components. An example of a loss of critical components is the reported loss of entire minicircle classes in *Leishmania tarentolae* during laboratory cultivation, which results in an inability to generate some correctly encoded genes. In the current work, we estimated the evolutionary rates of mitochondrial and nuclear genome regions of multiple natural *Leishmania* spp. We analyzed synonymous and non-synonymous substitutions and, rather unexpectedly, found that the coding regions of kinetoplast maxicircles are among the most variable regions of both genomes. In addition, we demonstrate that synonymous substitutions greatly predominate among maxicircle coding regions and that most maxicircle genes show signs of purifying selection. These results imply that maxicircles in natural *Leishmania* populations remain functional despite their high mutation rate.

Keywords: genome instability; *leishmania donovani*; *L. infantum*; *L. major*; *L. turanica*; SNP

Citation: Gerasimov, E.S.; Novozhilova, T.S.; Zimmer, S.L.; Yurchenko, V. Kinetoplast Genome of *Leishmania* spp. Is under Strong Purifying Selection. *Trop. Med. Infect. Dis.* **2023**, *8*, 384. https://doi.org/10.3390/tropicalmed8080384

Academic Editor: John Frean

Received: 28 June 2023
Revised: 24 July 2023
Accepted: 25 July 2023
Published: 27 July 2023

1. Introduction

Protists of the genus *Leishmania* (Kinetoplastida: Trypanosomatidae: Leishmaniinae) are important causative agents of leishmaniasis, a severe yet neglected human infectious ailment [1–3]. The disease has a wide range of clinical presentations, including cutaneous, mucocutaneous, and visceral forms [4,5]. Hundreds of millions of people in over 100 counties are at risk of *Leishmania* infection, with an estimated 1.5 million new cases annually [6].

Overall high levels of genome instability observed in many *Leishmania* species are an intriguing aspect of the parasite's biology and evolution [7]. Genomic instability is often cited as the key factor that allows *Leishmania* spp. to adapt to a wide range of vectors and vertebrate host species, spreading over dramatically different ecological niches [8,9]. Nuclear genome instability is probably linked to the absence of promoter-dependent gene expression regulation common for all trypanosomatids [10,11]. *Leishmania* DNA instability includes single-nucleotide polymorphisms as well as copy number variation of long genomic regions and even chromosomes (gene dosage) [9,12–15]. Both have been investigated in the context of identifying mechanisms of virulence and adaptation to hosts [8,16,17].

Because all previously mentioned studies focused on the nuclear genome only, we decided to look at the question of instability, including both the nuclear and mitochondrial (kinetoplast) genomes. Of note, the term kinetoplast refers to a prominent structure containing protein and the DNA of the *Leishmania* single mitochondrion. The kinetoplast genome consists of a few dozen maxicircle molecules (functional analogs of the mitochondrial genome of other eukaryotes) and a few thousand minicircle molecules (encoding small RNAs that guide a unique RNA editing process required for maturation of many maxicircle-encoded genes [18,19].

In the current study, we present an analysis of nuclear and kinetoplast genomes of strains of four closely related *Leishmania* (*Leishmania*) spp. (*L. donovani*, *L. infantum*, *L. major*, and *L. turanica*). Mutations, and specifically, point mutations, are the major playing field of selection and adaptation [20,21]. Using the rate of nuclear and kinetoplast single-nucleotide polymorphisms (SNPs) to estimate the degree of genetic divergence between strains of these four species, we show that their kinetoplast genomes are under strong purifying selection. Relatedly, we characterize the genes located in discrete, fast-evolving regions of *Leishmania* chromosomes.

2. Materials and Methods

2.1. Datasets

We used raw sequencing data of *Leishmania* strains belonging to four species: *L. turanica* (10 strains sequenced previously by our group [22], BioProject PRJNA888552); *L. major* (12 strains sequenced previously by our group [23], BioProject PRJNA763936); *L. donovani* and *L. infantum* (151 strains [24,25]). Each strain used in the analysis was isolated from a single individual. The accession numbers for sequencing reads used in the current study are summarized in Table S1 for the processed samples that passed quality and coverage controls (described below). All chosen datasets are Illumina paired-end WGS reads with similar read lengths of 100–150 bp.

2.2. Data Pre-Processing, Filtering, and Variant Calling

Nuclear genomes of *L. donovani* LV9 (TriTrypDB, release 62 [26]) and *L. major* Friedlin (TriTrypDB release 58) were used as references for variant calling. GenBank accession MN904518 and a contig from genome assembly of the strain Friedlin were used as maxicircle references for *L. donovani* and *L. major*, respectively. Sequencing reads were downloaded from SRA with 'fastq-dump', and quality control was performed with FastQC v. 6 [27] and MultiQC v. 1.14 [28] tools. Read trimming for adapter sequences and base quality analysis at 3' ends were performed with Trimmomatic v. 0.32 [29] in the paired-end mode, with minimal retained read length set to 75 bp. Read pairs were mapped onto respective references with Burrows–Wheeler Aligner v. 0.7.17 [30]. Alignment files were processed with SAMtools v. 1.17 [31,32]. Sequencing read libraries were retained for further analysis if they met all the following quality criteria: average maxicircle and nuclear genome coverage over 10×, at least 5 million reads mapped onto the reference genome, at least 25,000 reads containing the CSB3 (Conserved Sequence Block 3) sequence present on minicircles [33], and no coverage gaps of size over 10,000 bp. Those not meeting the criteria were discarded (Figure 1).

Variant calling was performed in BCFtools v. 1.17 [32] and FreeBayes v. 2 [34] with default parameters. Variant calling output files were processed with in-house Python scripts to account for minor formatting differences introduced by SNP callers. Custom Python scripts using scipy, numpy, and seaborn libraries were used for statistical testing and data visualization. Despite different overall numbers of variant calls, both variant calling programs returned very similar results after applying the full data processing pipeline. The distribution of nuclear genome SNPs per 18 kb window is shown in Figure S1 for the Freebayes dataset. The VCF files produced by BCFtools caller were further used to extract homozygous SNPs (AF > 0.85, coverage over 10, variant type = SNP) (Figure 1).

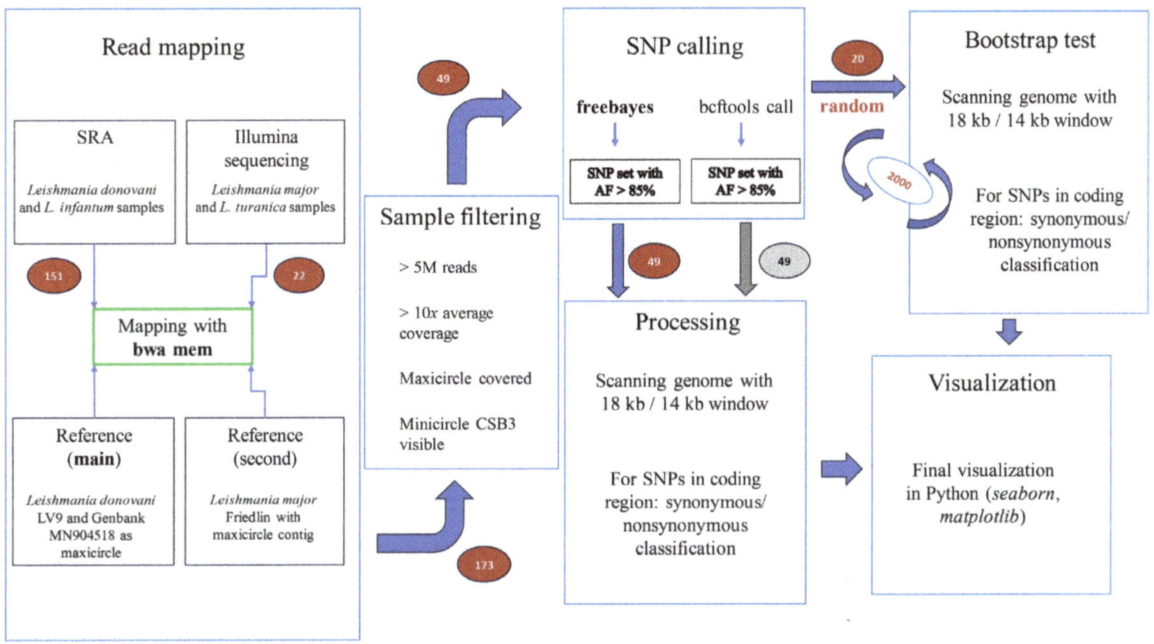

Figure 1. The scheme of the data processing pipeline depicting the main stages. The read mapping was performed for all 173 samples. All analyses were performed independently for both reference genome sequences. *Leishmania donovani* LV9 genome was marked as the 'MAIN' reference as all figures in the text are presented for this genome. *Leishmania major* Friedlin was used as the second reference genome. Number of samples (genomes) at each step is provided in red circles. Two SNP callers were used, and the output of both programs was processed identically downstream; blue arrow with number 49 indicates that Freebayes results were used for further presentation in the main text.

Classification of individual SNPs into categories of those contributing to either synonymous or non-synonymous amino acid substitutions was carried out with an in-house Python script. The program considered the possibility of indels and frameshift mutations, excluding genes with highly modified ORF sequences from further analysis. The script provided accurate per-site information for each annotated gene. Comparison of this script against the published snpEff tool v. 5.1 [35] using both nuclear and kinetoplast datasets showed almost identical results in terms of the total numbers of substitutions, with the snpEff reporting 1–3% more classified substitutions per sample.

2.3. Bootstrap Replications

For *L. donovani*, the median number of total synonymous and non-synonymous substitutions was calculated in an 18 kb non-overlapping window for the nuclear genome (average gene content 48.7%) and in and ~18 kb-long coding region (CR) of maxicircles (approximately 52% occupied by non-edited protein-coding genes [18]). Therefore, the size of the scanning window and coding capacity in these datasets is comparable for synonymous/non-synonymous SNP classification comparisons. For *L. major*, the window sizes were 14 kb (the length of its maxicircle CR). To estimate how robust nucleotide substitution median values are, we performed a subsampling (bootstrap), repeating our pipeline on 20 strain libraries randomly taken from a total of 49. We chose the subsample size of 20 to account for possible bias in *Leishmania donovani/infantum*- and *L. turanica/major*-prevalent subsamples to exclude the possible influence of unequal species representation on the median value.

2.4. Functional Annotation and Gene Enrichment

Genomic segment operations (gene and SNP intersection) were performed with Bed-Tools v. 2.31.0 [36]. Genes from variable genome loci were extracted from annotation using custom bash/Python scripts. Gene products were annotated using HMMScan in Pfam v. 35.0 [37] and NCBI-BLAST suite [38]. Gene ontology analysis was carried out with topGO v. 2.46.0 [39] and g:Profiler2 [40].

2.5. Selection Analysis

The CODEML program from the PAML package v. 4.8 [41] was used to estimate synonymous and non-synonymous rates (dS and dN). For the unedited and minimally edited maxicircle-encoded genes (*COI, COII, COIII, ND1, ND2, ND4, ND5, MURF2,* and *CYb*), the dN/dS ratio was calculated using M0 model estimation. The guiding tree for analysis was constructed from the full maxicircle CR alignment using FastTree v. 2.1 [42]. To reject possible evidence of positive selection, pairs of nested models M1a-M2a and M7-M8 were also compared with the likelihood-ratio test (LRT).

3. Results and Discussion

3.1. Whole Nuclear Genome Identification of Point Mutations

The first step in our analysis was to identify point mutations between individual strains of species *L. turanica, L. major, L. donovani* and *L. infantum,* and a *Leishmania* reference genome. As these four species are closely related [43,44], it would allow us to assess evolutionary changes on a relatively narrow scale of divergence. This will provide opportunities to address questions for which inter-species genetic variability is less suited to compare. These four species are highly syntenic and possess equal numbers of nuclear chromosomes (36). We thus identified strains of these species for which whole-genome sequencing (WGS) reads were available. Libraries for 173 different strains were identified. Out of these, 49 strains met the quality thresholds (See Section 2.2, Table S2: 12 *L. major*, 10 *L. turanica*, 14 *L. donovani*, and 13 *L. infantum*) that would allow SNP analysis of the nuclear and kinetoplast genomes, and the rest were excluded from the analysis. Libraries were mainly discarded due to low maxicircle coverage, indicating that commonly used WGS protocols typically result in low proportions of kinetoplast reads.

Each of the high-quality 49 strains was mapped to the well-annotated reference *L. donovani* LV9 scaffolds for SNP identification. General mapping statistics are presented in Table S2. The number of mapped reads for each strain library ranged from 8 to 56 million reads, with an average nuclear genome coverage between 23 and 225×. We consider this coverage level sufficient for homozygous SNPs calling [45]. An abnormally high percentage of read pairs were mapped onto different chromosomes for three strain libraries, indicative of chromosome translocations in these strains. This is especially pronounced in *L. infantum* IMT373cl1. The total number of homozygous SNPs detected for all 49 strain libraries is reported in Table S2. The percentage of nucleotide mismatch in the study ranged from 0.6 to 1.3% for strains of the more closely related *L. donovani/L. infantum* clade, and from 8 to 9.5% for *L. turanica/L. major* strains that are more distantly related to the reference strain to which they were mapped.

3.2. Analysis of Kinetoplast Maxicircle Genome Variation

The kinetoplast maxicircle of *Leishmania* spp. consists of the variable (divergent) and the coding region (DR and CR, respectively [46]). In the CR, which constitutes nearly half of the maxicircle [~16 of ~36 kb], genetic loci are closely spaced. Uniquely, these loci include traditional genes as well as cryptogenes that generate transcripts that undergo post-transcriptional editing to insert and delete uridines in order to encode their products [11,19]). As an annotated maxicircle sequence was not available to use as a reference scaffold on which to map individual strain reads, for this purpose, we manually annotated the 18 kb CR of *L. donovani* maxicircle (GenBank accession MN904518, sequence length 36,732 bp) with homology-based alignments with maxicircle genes of other *Leishmania* spp. The map is presented in Figure 2.

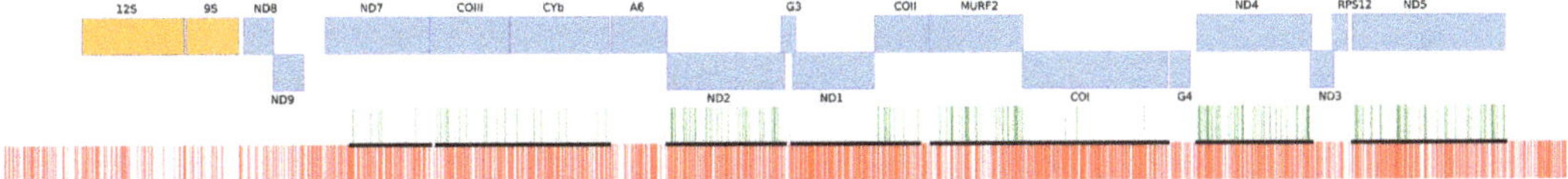

Figure 2. Distribution of SNPs over the *L. donovani* maxicircle CR. (**Top**) Map of *L. donovani* CR. The top and bottom boxes indicate genes encoded on different strands. Color indications for different types of genes: rRNA, orange; unedited genes and cryptogenes, blue. (**Bottom**) The red track shows the positions of all SNPs detected in 49 strains collectively. The green track shows the positions of non-synonymous SNPs (in locations where such a determination was possible). The black segments between tracks indicate maxicircle regions where the determination of synonymous or non-synonymous SNPs can be made. Namely, these are locations where ORFs can be deduced from the maxicircle sequence either because an entire mRNA is properly encoded or else a cryptogene possesses regions that are properly encoded and do not require editing.

Strain library reads were subsequently mapped to the newly annotated *L. donovani* CR reference sequence, with an average maxicircle CR coverage ranging from 13 to 318×.

The distribution of SNPs in the maxicircle CR is nonuniform (red track in Figure 2). We then asked whether these SNPs were synonymous or non-synonymous (green track in Figure 2). As it is not possible to assess the eventual functional outcome of SNPs within the edited regions of cryptogenes with DNA information only, these regions were left out of the analysis. A comparison of green and red tracks shows that the vast majority of SNPs in maxicircles are synonymous, approximately 5× more abundant than non-synonymous SNPs. The predominance of synonymous substitutions in maxicircle ORFs indicates that these loci are under strong purifying selection. Estimates of dN/dS ratios generated with CODEML gave values below 0.03, which implies a strong negative selection pattern. This supports amino acid sequence conservation of maxicircle genes as being vitally important for these *Leishmania* strains collected from the natural environments.

To test for the possibility that positive selection is acting on specific amino acids rather than full gene sequences, we also compared nested site models with CODEML. No selection was detected. Interestingly, *CYb* and *COI* genes are virtually depleted of non-synonymous substitutions, while in *ND2*, *MURF2*, *ND4*, and *ND5* genes, they are enriched relative to the other maxicircle ORF regions. This indicates that the evolutionary pressures on maxicircle gene sequences in *Leishmania* spp. are not uniform for maxicircle genes.

The importance of the various kinetoplast genome-encoded protein products in trypanosomatids is widely analyzed and debated. It is becoming increasingly clear that these parasites retain the capacity to remodel their maxicircles and/or maxicircle mRNA expression in order to adapt to their unique environments. For example, *Phytomonas* and *Vickermania* spp. lack genes encoding *CYb*, *COI*, *COII*, and *COIII* [47–51], apparently having circumvented the need for ATP generation through traditional aerobic respiration. Relatedly, during long laboratory cultivation, *Leishmania tarentolae* can lose groups of minicircles and thus their encoded regulatory RNAs that direct editing, but only those minicircles linked to the editing of specific maxicircle cryptogenes [52]. The maxicircle genes that are no longer able to be edited are likely not essential for *Leishmania* within the context of growth in a nutrient-rich and perturbation-free environment. However, SNP analysis of strains utilized here strongly suggests that the opposite is true in natural isolates of *Leishmania*: their maxicircles display signals of preference for amino acid sequence conservation (low proportion of non-synonymous SNPs), a sign of purifying selection.

3.3. Comparison of Nuclear Genome Variation with Maxicircle CR Region Genome Variation

We hypothesized that loci located on nuclear chromosomes have different mutation rates and are under different selection pressures than kinetoplast genome loci. To determine the extent of nuclear genome loci variation in a way that allows us to compare it with those of the maxicircle CR, we investigated synonymous to non-synonymous substitutions of ORFs within 18 kb windows of the nuclear genome, a window length comparable with the length of the maxicircle CR (Figure 3). The ratio of synonymous to non-synonymous substitutions in the nuclear genome is close to 1 (both values have similar distributions with means of 268 and 269 and median of 271 and 251, respectively). This sharply contrasts with the predominance of synonymous substitutions we found for maxicircle CR region loci (1038 synonymous versus 230 non-synonymous ones). Interestingly, the number of non-synonymous substitutions in maxicircles is located near the median value of that of 18 kb windows of the nuclear genome. The rate of amino acid substitutions due to SNPs in both genomes is actually similar. However, the DNA sequence of maxicircles is more likely to accumulate nucleotide substitutions over time (Figure 3A), as most maxicircle CR point mutations are synonymous (Figure 3C). We speculate that the kinetoplast DNA of *Leishmania* may be more prone to damage than its nuclear DNA due to mitochondrial processes that generate reactive oxygen species. Additionally, repair machinery that acts on the nuclear genome may ensure that fewer replication errors become nucleotide substitutions in later generations. In these circumstances in which maxicircle CR SNPs occur at a high rate, purifying selection acting on kinetoplast genes is the cleanest explanation for the modest rate of SNPs resulting in non-synonymous substitutions.

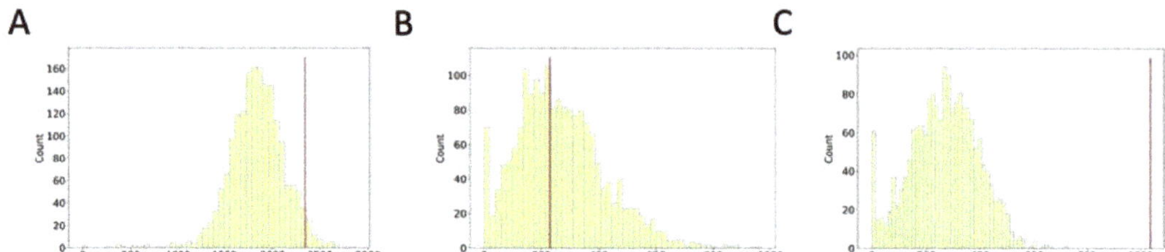

Figure 3. Distribution of nuclear genome SNPs per 18 kb window. Number of SNPs per window is given on the X axis, number of windows with that SNP count is given on Y axis. (**A**) For all SNPs; (**B**) for non-synonymous SNPs (including nonsense); (**C**) for synonymous SNPs. On all histograms, the red vertical line indicates the respective value for the kinetoplast maxicircle CR.

Our conclusion of strong purifying selective pressure on maxicircle coding loci in *Leishmania* is apparently not a phenomenon that applies to kinetoplast genomes in general. Analysis of the monoxenous trypanosomatids *Crithidia bombi* and *C. expoeki* parasitizing social bumblebees has shown that nucleotides of their kinetoplast genomes acquire fewer substitutions than their nuclear genomes [53]. In that work, SNP frequency was measured as the average number of nucleotides in which one SNP is encountered. If we apply the same approach to the current dataset, the averaged values for the *L. donovani*/*infantum* species complex will be 207 and 102 bp *per* SNP for nuclear and kinetoplast genomes, respectively. Thus, even when utilizing this alternative approach, we still conclude that the *Leishmania* strains analyzed acquire more nucleotide substitutions in their maxicircle CR than in their nuclear genome. We conclude that relative evolution rates for kinetoplast and nuclear genomes differ among members of Trypanosomatidae.

Considering that 80% of SNPs in *Leishmania* maxicircles are synonymous substitutions, it is interesting that a recent paper has reported the capacity for synonymous substitutions to be adaptive [54]. They can impact RNA folding and stability, modulate transcription and translation rates, or affect codon usage, which can influence the translation rate. The latter is especially important for trypanosomatid maxicircles lacking identifiable transcription regulatory elements.

3.4. Robustness of the Results

The results presented here are not likely to be the result of reference genome bias. When the genome assembly of *L. major* Friedlin rather than that of *L. donovani* was used as a reference, similar numbers of SNPs were obtained and nuclear to maxicircle CR SNP ratios were similar (Figure S2 and Table S2). To demonstrate the robustness of the results with regard to strain library selection, we performed 2000 bootstrap replications, randomly choosing 20 samples out of the pool of 49 and repeated SNP identification (using *L. donovani* as reference). For each replication, a median number of SNPs in each 18 kb window for the nuclear genome and the number of SNPs in maxicircle CR was calculated. The distribution of these two values for 2000 bootstrap replicates is shown in Figure S3. Our results demonstrate that independent of the strain libraries utilized for analysis, the median SNP numbers for nuclear genome 18 kb windows are lower than that of the maxicircle CR. As expected, with replicates that utilized only 20 strain libraries, the absolute value of the median number of SNPs is less: by decreasing the sample size from 49 to 20 strain libraries, the median number of SNPs shifted from 1836 to 1733 (~8%).

3.5. Analysis of Highly Variable Nuclear Genome Loci

As can be surmised by Figure 3A, there is variability in the number of SNPs in different nuclear genomic regions. To determine where genomic regions with high nucleotide substitution rates are positioned in the *Leishmania* nuclear genome, we generated chromosome maps on which we plotted the 18 kb genomic windows with total SNP numbers higher than that of maxicircle CRs (Figure 4). These highly variable regions are distributed on the chromosomes in a non-random fashion. Some chromosomes were free of these regions, and others were densely covered. For example, genetically variable regions occupy almost 30% of Chromosome 31. This chromosome is tetra(poly)-somic in most (if not all) *Leishmania* spp. analyzed and is likely involved in recombination processes [55–57]. In contrast, we found no direct connection between high SNP-containing windows and chromosome telomeres, centromeres, or repeat-rich regions. The same analysis using the *L. major* genome as a reference resulted in similar distributions with just a few minor differences, most probably caused by fluctuations in window size selection (Figure S4).

Our results show a picture of non-randomly distributed point mutations across the *Leishmania* nuclear genome. Recent findings suggest that genome instability plays an important role in *Leishmania* evolution as it is connected with gene dosage changes that were also shown to be non-randomly distributed across the genome and highly reproducible in populations [7]. Clearly, copy number variations (CNVs) are under selection. Various kinds of repeats and low-complexity regions may act as the drivers for gene dosage through a variety of mechanisms [58]. In *Leishmania* spp. genomes, the association of repeats and CNVs is statistically significant [7]. Conversely, maxicircle CRs lack intergenic regions and, thus, the expression of their genes cannot be regulated by the same mechanism. As discussed above, the synonymous changes we identified have the capacity to modulate translation and, as such, may be adaptive. Based on that, we can hypothesize that the genetically variable regions of the nuclear genome and the maxicircle CR display the same pattern of adaptive evolution.

Finally, we extracted the genes located in the 18 kb windows with high levels of nucleotide polymorphisms and analyzed the likely function of their products to determine if any gene category was enriched among the extracted genes. Homology-based annotations using UniProt and Pfam databases against other annotated trypanosomatid genomes revealed large numbers of genes encoding mitochondria-imported proteins, as well as genes encoding proteins involved in RNA modification and binding, core proteins involved in cell division and mobility, components of the ribosome, and several chaperones. However, formal gene ontology enrichment analysis did not show any significantly enriched categories. Therefore, it is less likely that the identity of the proteins encoded in the regions of the nuclear genome with high nucleotide variability is driving their presence.

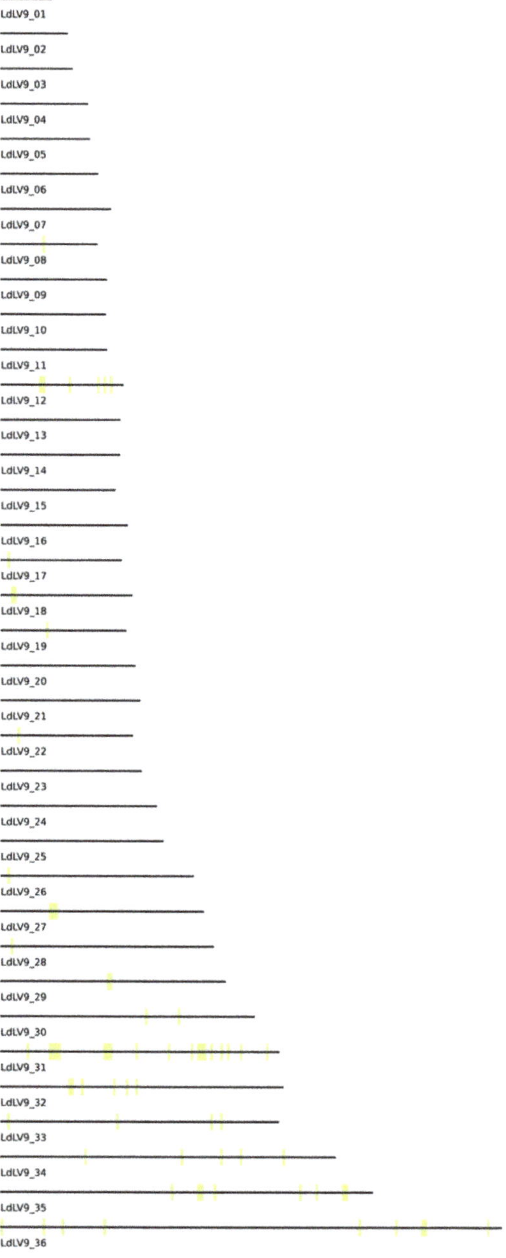

Figure 4. Genomic layout of 18 kb windows with per nucleotide substitution rate equal or higher than that in maxicircle CR. All 36 chromosomes are scaled proportionally to their length.

4. Conclusions

In this work, we investigated nucleotide polymorphisms among 49 *Leishmania* strains and the evolutionary implications of where and in what abundance they were found.

The maxicircle CR displays high nucleotide variability, with more SNPs on average than the nuclear genome. This is unexpected and counterintuitive, provided the important functional role the products of maxicircle genes play in trypanosomatid biochemistry. The ratio of maxicircle CR non-synonymous to synonymous substitutions is very low (<0.2), indicating strong purifying selection acting on the *Leishmania* genes encoded in the kinetoplast genome. These results are strong evidence that natural populations of *Leishmania* spp. retain the capacity and a necessity to express functional kinetoplast genes. As laboratory cultures of trypanosomatids can lose their ability to functionally express some kinetoplast genes, these results reveal potential limitations of culture experiments. This is particularly true for the investigation of kinetoplast DNA expression and RNA editing, as these processes initiate from a genome template that may be altered in continuously grown laboratory cultures. Nevertheless, the result obtained for 49 strains of just two species of *Leishmania* must be taken with caution and additional work on other species of Trypanosomatidae (including other species and isolates of *Leishmania*) is warranted in the future. It may also be important to focus attention on the monoxenous trypanosomatids and mutants of the dixenous ones with alterations in their life cycle in order to understand the extent of purifying selection on maxicircles in these species. On a more general note, one way to answer the question of what might affect the high occurrence of synonymous codons is to determine how often these synonymous mutations contribute to adaptation in natural settings beyond the highly contrived laboratory conditions. In other words, does fitness in vitro predict the prevalence of a mutation across a phylogeny? This is a relatively unexplored field, but it was proposed that differential usage of synonymous mRNA variants due to variable speed of decoding on ribosome may be a sign of a "code within the genetic code" for guiding accurate co-translational protein folding and subsequently maintenance of cellular functions [59]. More population-scale sequencing of Trypanosomatidae is needed to address this and related questions [60].

Supplementary Materials: The following supporting information can be downloaded at: https://www.mdpi.com/article/10.3390/tropicalmed8080384/s1, Figure S1: Distribution of SNPs per 18 kb window for nuclear genome of *L. donovani* as reference genome and Freebayes as SNP-caller. The vertical red line denotes the number of SNPs detected by Freebayes in 18 kb maxicircle coding region with flanks. Figure S2: Distribution of SNPs per 14 kb window for the nuclear genome using *L. major* as a reference genome. The vertical red line is the number of SNPs in the *L. major* maxicircle CR; Figure S3: Distribution of the median number of SNPs per 18 kb window for the nuclear genome (using *L. donovani* reference) for 2000 bootstrap replications (sample size 20) is shown in green, and the SNPs median number per maxicircle CR is shown in orange. Green and red vertical lines are medians of these distributions, respectively. Black vertical lines show the median SNP numbers for the nuclear (left line) and maxicircle CR (right line) when all 49 strain libraries are used; Figure S4: Genomic layout of 14 kb windows with *per* nucleotide substitution rate equal or higher than that of maxicircle CR (genomic reference: *L. major*). All 36 chromosomes are scaled proportionally to their length; Table S1: Accession numbers for sequencing reads used in current study; Table S2: General mapping statistics for 49 *Leishmania* strains onto *L. donovani* and *L. major* reference genomes.

Author Contributions: Conceptualization, E.S.G. and V.Y.; methodology, E.S.G.; validation, T.S.N. and S.L.Z.; investigation, E.S.G., T.S.N., S.L.Z. and V.Y.; resources, E.S.G.; data curation, T.S.N. and S.L.Z.; writing, original draft preparation, E.S.G., writing, review and editing, all authors.; visualization, E.S.G. and T.S.N.; funding acquisition, E.S.G. and V.Y. All authors have read and agreed to the published version of the manuscript.

Funding: This research was funded by the Russian Science Foundation, grant number RSF 19-74-10008 to E.S.G. The computational infrastructure used in analysis was purchased using resources provided by the European Regional Funds CZ.02.1.01/16_019/0000759 to V.Y. The funders had no role in the design of the study; in the collection, analyses, or interpretation of data; in the writing of the manuscript; or in the decision to publish the results.

Institutional Review Board Statement: Not applicable.

Informed Consent Statement: Not applicable.

Data Availability Statement: Publicly available datasets were analyzed in this study. These data can be found here: https://www.ncbi.nlm.nih.gov/ (accessed on 15 June 2023).

Acknowledgments: We thank members of our laboratories for stimulating discussions.

Conflicts of Interest: V.Y. is a Section Editor-in-Chief for the Tropical Medicine and Infectious Disease. Other authors declare no conflict of interest.

References

1. Burza, S.; Croft, S.L.; Boelaert, M. Leishmaniasis. *Lancet* **2018**, *392*, 951–970. [CrossRef] [PubMed]
2. Steverding, D. The history of leishmaniasis. *Parasites Vectors* **2017**, *10*, 82. [CrossRef] [PubMed]
3. Kostygov, A.Y.; Karnkowska, A.; Votýpka, J.; Tashyreva, D.; Maciszewski, K.; Yurchenko, V.; Lukeš, J. Euglenozoa: Taxonomy, diversity and ecology, symbioses and viruses. *Open Biol.* **2021**, *11*, 200407. [CrossRef] [PubMed]
4. Bruschi, F.; Gradoni, L. *The Leishmaniases: Old Neglected Tropical Diseases*; Springer: Cham, Switzerland, 2018; 245p.
5. WHO. Leishmaniasis. 2023. Available online: https://www.who.int/en/news-room/fact-sheets/detail/leishmaniasis (accessed on 1 June 2023).
6. Alvar, J.; Velez, I.D.; Bern, C.; Herrero, M.; Desjeux, P.; Cano, J.; Jannin, J.; den Boer, M. Leishmaniasis worldwide and global estimates of its incidence. *PLoS ONE* **2012**, *7*, e35671.
7. Bussotti, G.; Piel, L.; Pescher, P.; Domagalska, M.A.; Rajan, K.S.; Cohen-Chalamish, S.; Doniger, T.; Hiregange, D.G.; Myler, P.J.; Unger, R.; et al. Genome instability drives epistatic adaptation in the human pathogen *Leishmania*. *Proc. Natl. Acad. Sci. USA* **2021**, *118*, e2113744118. [CrossRef]
8. Bussotti, G.; Gouzelou, E.; Cortes Boite, M.; Kherachi, I.; Harrat, Z.; Eddaikra, N.; Mottram, J.C.; Antoniou, M.; Christodoulou, V.; Bali, A.; et al. *Leishmania* genome dynamics during environmental adaptation reveal strain-specific differences in gene copy number variation, karyotype instability, and telomeric amplification. *mBio* **2018**, *9*, e01399-01318. [CrossRef]
9. Dumetz, F.; Imamura, H.; Sanders, M.; Seblová, V.; Myšková, J.; Pescher, P.; Vanaerschot, M.; Meehan, C.J.; Cuypers, B.; De Muylder, G.; et al. Modulation of aneuploidy in *Leishmania donovani* during adaptation to different in vitro and in vivo environments and its impact on gene expression. *mBio* **2017**, *8*, e00599-00517. [CrossRef]
10. Clayton, C. Regulation of gene expression in trypanosomatids: Living with polycistronic transcription. *Open Biol.* **2019**, *9*, 190072. [CrossRef]
11. Maslov, D.A.; Opperdoes, F.R.; Kostygov, A.Y.; Hashimi, H.; Lukeš, J.; Yurchenko, V. Recent advances in trypanosomatid research: Genome organization, expression, metabolism, taxonomy and evolution. *Parasitology* **2019**, *146*, 1–27. [CrossRef]
12. Iantorno, S.A.; Durrant, C.; Khan, A.; Sanders, M.J.; Beverley, S.M.; Warren, W.C.; Berriman, M.; Sacks, D.L.; Cotton, J.A.; Grigg, M.E. Gene expression in *Leishmania* is regulated predominantly by gene dosage. *mBio* **2017**, *8*, e01393-01317. [CrossRef]
13. Leprohon, P.; Légaré, D.; Raymond, F.; Madore, E.; Hardiman, G.; Corbeil, J.; Ouellette, M. Gene expression modulation is associated with gene amplification, supernumerary chromosomes and chromosome loss in antimony-resistant *Leishmania infantum*. *Nucleic Acids Res.* **2009**, *37*, 1387–1399. [CrossRef] [PubMed]
14. Rogers, M.B.; Hilley, J.D.; Dickens, N.J.; Wilkes, J.; Bates, P.A.; Depledge, D.P.; Harris, D.; Her, Y.; Herzyk, P.; Imamura, H.; et al. Chromosome and gene copy number variation allow major structural change between species and strains of *Leishmania*. *Genome Res.* **2011**, *21*, 2129–2142. [CrossRef]
15. Prieto Barja, P.; Pescher, P.; Bussotti, G.; Dumetz, F.; Imamura, H.; Kedra, D.; Domagalska, M.; Chaumeau, V.; Himmelbauer, H.; Pages, M.; et al. Haplotype selection as an adaptive mechanism in the protozoan pathogen *Leishmania donovani*. *Nat. Ecol. Evol.* **2017**, *1*, 1961–1969. [CrossRef]
16. Tosi, L.R.O.; Denny, P.W.; De Oliveira, C.I.; Damasceno, J.D. *Leishmania* genome variability: Impacts on parasite evolution, parasitism and leishmaniases control. *Front. Cell. Infect. Microbiol.* **2023**, *13*, 1171962. [CrossRef] [PubMed]
17. Santi, A.M.M.; Murta, S.M.F. Impact of genetic diversity and genome plasticity of *Leishmania* spp. in treatment and the search for novel chemotherapeutic targets. *Front. Cell. Infect. Microbiol.* **2022**, *12*, 826287. [CrossRef]
18. Aphasizheva, I.; Alfonzo, J.; Carnes, J.; Cestari, I.; Cruz-Reyes, J.; Goringer, H.U.; Hajduk, S.; Lukeš, J.; Madison-Antenucci, S.; Maslov, D.A.; et al. Lexis and grammar of mitochondrial RNA processing in trypanosomes. *Trends Parasitol.* **2020**, *36*, 337–355. [CrossRef]
19. Read, L.K.; Lukeš, J.; Hashimi, H. Trypanosome RNA editing: The complexity of getting U in and taking U out. *Wiley Interdiscip. Rev. RNA* **2016**, *7*, 33–51. [CrossRef]
20. Lynch, M. Evolution of the mutation rate. *Trends Genet.* **2010**, *26*, 345–352. [CrossRef]
21. Sniegowski, P.D.; Gerrish, P.J.; Johnson, T.; Shaver, A. The evolution of mutation rates: Separating causes from consequences. *Bioessays* **2000**, *22*, 1057–1066. [CrossRef]
22. Novozhilova, T.S.; Chistyakov, D.S.; Akhmadishina, L.V.; Lukashev, A.N.; Gerasimov, E.S.; Yurchenko, V. Genomic analysis of *Leishmania turanica* strains from different regions of Central Asia. *PLoS Negl. Trop. Dis.* **2023**, *17*, e0011145. [CrossRef]
23. Kostygov, A.Y.; Grybchuk, D.; Kleschenko, Y.; Chistyakov, D.S.; Lukashev, A.N.; Gerasimov, E.S.; Yurchenko, V. Analyses of *Leishmania*-LRV co-phylogenetic patterns and evolutionary variability of viral proteins. *Viruses* **2021**, *13*, 2305. [CrossRef] [PubMed]

24. Franssen, S.U.; Durrant, C.; Stark, O.; Moser, B.; Downing, T.; Imamura, H.; Dujardin, J.C.; Sanders, M.J.; Mauricio, I.; Miles, M.A.; et al. Global genome diversity of the *Leishmania donovani* complex. *eLife* **2020**, *9*, e51243. [CrossRef] [PubMed]
25. Imamura, H.; Downing, T.; van den Broeck, F.; Sanders, M.J.; Rijal, S.; Sundar, S.; Mannaert, A.; Vanaerschot, M.; Berg, M.; De Muylder, G.; et al. Evolutionary genomics of epidemic visceral leishmaniasis in the Indian subcontinent. *eLife* **2016**, *5*, e12613. [CrossRef] [PubMed]
26. Aslett, M.; Aurrecoechea, C.; Berriman, M.; Brestelli, J.; Brunk, B.P.; Carrington, M.; Depledge, D.P.; Fischer, S.; Gajria, B.; Gao, X.; et al. TriTrypDB: A functional genomic resource for the Trypanosomatidae. *Nucleic Acids Res.* **2010**, *38*, D457–D462. [CrossRef]
27. Andrews, S. FastQC: A Quality Control Tool for High Throughput Sequence Data. 2019. Available online: http://www.bioinformatics.babraham.ac.uk/projects/fastqc (accessed on 27 August 2022).
28. Ewels, P.; Magnusson, M.; Lundin, S.; Käller, M. MultiQC: Summarize analysis results for multiple tools and samples in a single report. *Bioinformatics* **2016**, *32*, 3047–3048. [CrossRef] [PubMed]
29. Bolger, A.M.; Lohse, M.; Usadel, B. Trimmomatic: A flexible trimmer for Illumina sequence data. *Bioinformatics* **2014**, *30*, 2114–2120. [CrossRef]
30. Li, H.; Durbin, R. Fast and accurate short read alignment with Burrows-Wheeler transform. *Bioinformatics* **2009**, *25*, 1754–1760. [CrossRef]
31. Li, H.; Handsaker, B.; Wysoker, A.; Fennell, T.; Ruan, J.; Homer, N.; Marth, G.; Abecasis, G.; Durbin, R. Genome project data processing, S. The Sequence Alignment/Map format and SAMtools. *Bioinformatics* **2009**, *25*, 2078–2079. [CrossRef]
32. Danecek, P.; Bonfield, J.K.; Liddle, J.; Marshall, J.; Ohan, V.; Pollard, M.O.; Whitwham, A.; Keane, T.; McCarthy, S.A.; Davies, R.M.; et al. Twelve years of SAMtools and BCFtools. *Gigascience* **2021**, *10*, giab008. [CrossRef]
33. Yurchenko, V.; Kolesnikov, A.A.; Lukeš, J. Phylogenetic analysis of Trypanosomatina (Protozoa: Kinetoplastida) based on minicircle conserved regions. *Folia Parasitol.* **2000**, *47*, 1–5. [CrossRef]
34. Garrison, E.; Marth, G. Haplotype-based variant detection from short-read sequencing. *arXiv* **2012**, arXiv:1207.3907.
35. Cingolani, P.; Platts, A.; Wang le, L.; Coon, M.; Nguyen, T.; Wang, L.; Land, S.J.; Lu, X.; Ruden, D.M. A program for annotating and predicting the effects of single nucleotide polymorphisms, SnpEff: SNPs in the genome of *Drosophila melanogaster* strain w1118; iso-2; iso-3. *Fly* **2012**, *6*, 80–92. [CrossRef]
36. Quinlan, A.R. BEDTools: The swiss-army tool for genome feature analysis. *Curr. Protoc. Bioinform.* **2014**, *47*, 11.12.11-11.12.34. [CrossRef] [PubMed]
37. Mistry, J.; Chuguransky, S.; Williams, L.; Qureshi, M.; Salazar, G.A.; Sonnhammer, E.L.L.; Tosatto, S.C.E.; Paladin, L.; Raj, S.; Richardson, L.J.; et al. Pfam: The protein families database in 2021. *Nucleic Acids Res.* **2021**, *49*, D412–D419. [CrossRef] [PubMed]
38. Johnson, M.; Zaretskaya, I.; Raytselis, Y.; Merezhuk, Y.; McGinnis, S.; Madden, T.L. NCBI BLAST: A better web interface. *Nucleic Acids Res.* **2008**, *36*, W5–W9. [CrossRef]
39. Alexa, A.; Rahnenfuhrer, J. *topGO: Enrichment Analysis for Gene Ontology*; R Package; Bioconductor: Boston, MA, USA, 2021.
40. Kolberg, L.; Raudvere, U.; Kuzmin, I.; Vilo, J.; Peterson, H. gprofiler2—An R package for gene list functional enrichment analysis and namespace conversion toolset g:Profiler. *F1000Res* **2020**, *9*, 709. [CrossRef]
41. Yang, Z. PAML 4: Phylogenetic analysis by maximum likelihood. *Mol. Biol. Evol.* **2007**, *24*, 1586–1591. [CrossRef]
42. Price, M.N.; Dehal, P.S.; Arkin, A.P. FastTree 2—Approximately maximum-likelihood trees for large alignments. *PLoS ONE* **2010**, *5*, e9490. [CrossRef]
43. Albanaz, A.T.S.; Gerasimov, E.S.; Shaw, J.J.; Sádlová, J.; Lukeš, J.; Volf, P.; Opperdoes, F.R.; Kostygov, A.Y.; Butenko, A.; Yurchenko, V. Genome analysis of *Endotrypanum* and *Porcisia* spp., closest phylogenetic relatives of *Leishmania*, highlights the role of amastins in shaping pathogenicity. *Genes* **2021**, *12*, 444. [CrossRef]
44. Espinosa, O.A.; Serrano, M.G.; Camargo, E.P.; Teixeira, M.M.G.; Shaw, J.J. An appraisal of the taxonomy and nomenclature of trypanosomatids presently classified as *Leishmania* and *Endotrypanum*. *Parasitology* **2018**, *145*, 430–442. [CrossRef]
45. Liu, J.; Shen, Q.; Bao, H. Comparison of seven SNP calling pipelines for the next-generation sequencing data of chickens. *PLoS ONE* **2022**, *17*, e0262574. [CrossRef]
46. Gerasimov, E.S.; Zamyatnina, K.A.; Matveeva, N.S.; Rudenskaya, Y.A.; Kraeva, N.; Kolesnikov, A.A.; Yurchenko, V. Common structural patterns in the maxicircle divergent region of Trypanosomatidae. *Pathogens* **2020**, *9*, 100. [CrossRef]
47. Opperdoes, F.R.; Butenko, A.; Zakharova, A.; Gerasimov, E.S.; Zimmer, S.L.; Lukeš, J.; Yurchenko, V. The remarkable metabolism of *Vickermania ingenoplastis*: Genomic predictions. *Pathogens* **2021**, *10*, 68. [CrossRef]
48. Maslov, D.A.; Nawathean, P.; Scheel, J. Partial kinetoplast-mitochondrial gene organization and expression in the respiratory deficient plant trypanosomatid *Phytomonas serpens*. *Mol. Biochem. Parasitol.* **1999**, *99*, 207–221. [CrossRef] [PubMed]
49. Verner, Z.; Čermáková, P.; Škodová, I.; Kováčová, B.; Lukeš, J.; Horváth, A. Comparative analysis of respiratory chain and oxidative phosphorylation in *Leishmania tarentolae*, *Crithidia fasciculata*, *Phytomonas serpens* and procyclic stage of *Trypanosoma brucei*. *Mol. Biochem. Parasitol.* **2014**, *193*, 55–65. [CrossRef]
50. Porcel, B.M.; Denoeud, F.; Opperdoes, F.R.; Noel, B.; Madoui, M.-A.; Hammarton, T.C.; Field, M.C.; Da Silva, C.; Couloux, A.; Poulain, J.; et al. The streamlined genome of *Phytomonas* spp. relative to human pathogenic kinetoplastids reveals a parasite tailored for plants. *PLoS Genet.* **2014**, *10*, e1004007. [CrossRef]
51. Čermáková, P.; Maďarová, A.; Baráth, P.; Bellová, J.; Yurchenko, V.; Horváth, A. Differences in mitochondrial NADH dehydrogenase activities in trypanosomatids. *Parasitology* **2021**, *148*, 1161–1170. [CrossRef] [PubMed]

52. Thiemann, O.H.; Maslov, D.A.; Simpson, L. Disruption of RNA editing in *Leishmania tarentolae* by the loss of minicircle-encoded guide RNA genes. *EMBO J.* **1994**, *13*, 5689–5700. [CrossRef] [PubMed]
53. Gerasimov, E.; Zemp, N.; Schmid-Hempel, R.; Schmid-Hempel, P.; Yurchenko, V. Genomic variation among strains of *Crithidia bombi* and *C. expoeki*. *mSphere* **2019**, *4*, e00482-00419. [CrossRef] [PubMed]
54. Bailey, S.F.; Alonso Morales, L.A.; Kassen, R. Effects of synonymous mutations beyond codon bias: The evidence for adaptive synonymous substitutions from microbial evolution experiments. *Genome Biol. Evol.* **2021**, *13*, evab141. [CrossRef]
55. Downing, T.; Imamura, H.; Decuypere, S.; Clark, T.G.; Coombs, G.H.; Cotton, J.A.; Hilley, J.D.; de Doncker, S.; Maes, I.; Mottram, J.C.; et al. Whole genome sequencing of multiple *Leishmania donovani* clinical isolates provides insights into population structure and mechanisms of drug resistance. *Genome Res.* **2011**, *21*, 2143–2156. [CrossRef]
56. Glans, H.; Lind Karlberg, M.; Advani, R.; Bradley, M.; Alm, E.; Andersson, B.; Downing, T. High genome plasticity and frequent genetic exchange in *Leishmania tropica* isolates from Afghanistan, Iran and Syria. *PLoS Negl. Trop. Dis.* **2021**, *15*, e0010110. [CrossRef] [PubMed]
57. Fiebig, M.; Kelly, S.; Gluenz, E. Comparative life cycle transcriptomics revises *Leishmania mexicana* genome annotation and links a chromosome duplication with parasitism of vertebrates. *PLoS Pathog.* **2015**, *11*, e1005186. [CrossRef]
58. Hastings, P.J.; Lupski, J.R.; Rosenberg, S.M.; Ira, G. Mechanisms of change in gene copy number. *Nat. Rev. Genet.* **2009**, *10*, 551–564. [CrossRef] [PubMed]
59. Mitra, S.; Ray, S.K.; Banerjee, R. Synonymous codons influencing gene expression in organisms. *Res. Rep. Biochem.* **2016**, *6*, 57–65. [CrossRef]
60. Yurchenko, V.; Butenko, A.; Kostygov, A.Y. Genomics of Trypanosomatidae: Where we stand and what needs to be done? *Pathogens* **2021**, *10*, 1124. [CrossRef] [PubMed]

Tropical Medicine and Infectious Disease

MDPI

Review

Detection of Tropical Diseases Caused by Mosquitoes Using CRISPR-Based Biosensors

Salma Nur Zakiyyah [1], Abdullahi Umar Ibrahim [2], Manal Salah Babiker [3], Shabarni Gaffar [1], Mehmet Ozsoz [2], Muhammad Ihda H. L. Zein [1] and Yeni Wahyuni Hartati [1,*]

[1] Department of Chemistry, Faculty of Mathematics and Natural Sciences, Universitas Padjadjaran, Sumedang 45363, Indonesia
[2] Department of Biomedical Engineering, Near East University, Mersin 10, 99010 Nicosia, Turkey
[3] Department of Medical Biology and Genetics, Near East University, Mersin 10, 99010 Nicosia, Turkey
*   Correspondence: yeni.w.hartati@unpad.ac.id

Abstract: Tropical diseases (TDs) are among the leading cause of mortality and fatality globally. The emergence and reemergence of TDs continue to challenge healthcare system. Several tropical diseases such as yellow fever, tuberculosis, cholera, Ebola, HIV, rotavirus, dengue, and malaria outbreaks have led to endemics and epidemics around the world, resulting in millions of deaths. The increase in climate change, migration and urbanization, overcrowding, and other factors continue to increase the spread of TDs. More cases of TDs are recorded as a result of substandard health care systems and lack of access to clean water and food. Early diagnosis of these diseases is crucial for treatment and control. Despite the advancement and development of numerous diagnosis assays, the healthcare system is still hindered by many challenges which include low sensitivity, specificity, the need of trained pathologists, the use of chemicals and a lack of point of care (POC) diagnostic. In order to address these issues, scientists have adopted the use of CRISPR/Cas systems which are gene editing technologies that mimic bacterial immune pathways. Recent advances in CRISPR-based biotechnology have significantly expanded the development of biomolecular sensors for diagnosing diseases and understanding cellular signaling pathways. The CRISPR/Cas strategy plays an excellent role in the field of biosensors. The latest developments are evolving with the specific use of CRISPR, which aims for a fast and accurate sensor system. Thus, the aim of this review is to provide concise knowledge on TDs associated with mosquitoes in terms of pathology and epidemiology as well as background knowledge on CRISPR in prokaryotes and eukaryotes. Moreover, the study overviews the application of the CRISPR/Cas system for detection of TDs associated with mosquitoes.

Keywords: CRISPR; electrochemistry; biosensor; tropical disease

Citation: Zakiyyah, S.N.; Ibrahim, A.U.; Babiker, M.S.; Gaffar, S.; Ozsoz, M.; Zein, M.I.H.L.; Hartati, Y.W. Detection of Tropical Diseases Caused by Mosquitoes Using CRISPR-Based Biosensors. *Trop. Med. Infect. Dis.* **2022**, *7*, 309. https://doi.org/10.3390/tropicalmed7100309

Academic Editors: Vyacheslav Yurchenko and Thomas Dorlo

Received: 22 August 2022
Accepted: 11 October 2022
Published: 17 October 2022

Publisher's Note: MDPI stays neutral with regard to jurisdictional claims in published maps and institutional affiliations.

1. Introduction

The world is constantly facing the outbreak and reemergence of tropical diseases (TDs). The history of TDs dates back to ancient times, including the Roman and Egyptian empires. TDs are defined as diseases that are prevalent or indigenous to tropical and subtropical regions. Some of the most common TDs include malaria, cholera, yellow fever, dengue, Zika, etc. [1,2].

Mosquitoes are arthropods which are involved in the transmission of multiple pathogens, causing diseases that include dengue fever, chikungunya fever, malaria, filariasis, *Japanese encephalitis*, Zika, etc. TDs caused by mosquitoes have been associated with the mortality of humans every year. Therefore, mosquitoes are a major public health threat and thus can affect the economies of infected regions or countries. In the past few years, synthetic pesticides have been used to control mosquitoes. However, synthetic pesticides can cause contamination, kill many beneficial insects, and lead to the development of resistant-types after long-term use [3].

Advances in microscopy, molecular biology, biochemistry, diagnostic techniques, and treatment approaches in the 20th and 21st centuries have contributed to the understanding of the genetic contents of pathogens, biochemical reactions, and controlled such as susceptibility and resistance to drugs, vaccination of tropical diseases. The use of computer-aided techniques and other relevant technologies continue to aid experts in mapping regions of origins, predictions of spread, and creation of awareness using media outlets [2,4].

Early diagnosis of TDs is crucial for timely treatment, increasing patients' survival rates, preventing further outbreaks, and minimizing the cost of diagnosis. Advances in science and technology continue to improve the accuracy, sensitivity, and specificity of diagnostic approaches. Currently, molecular testing and antibody-based approaches are regarded as the standard approaches for diagnosing TDs. Other techniques used by medical experts include imaging approaches (such as X-ray, CT scans, ultrasound), blood tests, microscopy, sputum tests, etc. These techniques have several limitations which include the need for sophisticated devices, the need for highly trained and skilled pathologists and medical laboratory technicians, the use of toxic chemical reagents, and a lack of POC diagnostics [4].

Genome editing technology is regarded as one solution that can be used to modify the genome of organisms and harness their mechanism as a form of biomimetic approach for accurate detection of diseases. The three most widely techniques employed for manipulating the genomes of different species, including mosquitoes, include Zinc-finger Nucleases (ZFNs) and Transcription Activator-like Effector Nucleases (TALENs). Advancement in molecular biology has led to the discovery of unique immune pathways utilized by bacteria to fight against viruses such as bacteriophages [5,6]. This immune response approach is termed CRISPR (Clustered Regularly Interspaced Short Palindromic Repeat). When viruses invade bacteria, they inject their nucleic acid (in the form of RNA or DNA) which hijack the bacterial DNA replication system, generating more viruses and subsequently destroying the bacterial cell. To prevent this type of invasion, bacteria utilize a three-step process that includes adaptation, expression (biogenesis or recognition), and interference to ensure immunity [7].

Since the first deployment of CRISPR/Cas9 for genome editing in 2012, many researchers have successfully applied this technique to accurately edit the genomes of a variety of organisms such as bacteria, yeast, plants, and animals including mosquitoes. CRISPR/Cas technology is cheaper, easy-to-use, and accurate compared to ZFNs and TALENs. Scientists biomimic this pathway by designing a synthetic RNA known as single guide RNA (SgRNA) which binds with the target (pathogen) nucleic acid (RNA in viruses and DNA in bacteria and parasites). The application of the CRISPR/Cas system has proven to be among the most, reliable, accurate, sensitive, specific, and fast methods for screening pathogens associated with TDs. CRISPR/Cas9 genome editing technology also enables the modification of the target genes of pests. This is especially useful in controlling vector-borne diseases caused by mosquitoes [3,6,8].

Scope and Contribution

Throughout history, TDs have caused health issues around the world and contributed significantly to the mortality rate as well as the socioeconomic status of infected regions. Among numerous TDs, diseases associated with mosquitoes as carriers are the most common types of TDs with a high incidence and death toll. Despite measures taken by international organization such as the WHO, UNICEF, Red Cross, etc., to mitigate the array of infections cause by mosquitoes, progress is hindered by several factors such as resistance to insecticide, the drug resistance of parasites, climate change, urbanization, deforestation, lack of awareness, and lack of standard medical resources and social amenities in rural and underdeveloped regions. Early diagnosis is the first line of action in terms of the management of disease. In recent years, scientists have developed several laboratory assays for the rapid screening and detection of pathogens. Despite the progress achieved, current approaches are still hindered by factors such as low sensitivity, specificity, accuracy, false positive results, and misdiagnosis. The discovery of CRISPR/Cas systems in prokaryotes has open the window for scientists to repurpose or biomimic this approach in living cells. The CRISPR toolbox is a Pandora's box that has several applications which include genetic modification (knock-in and knock-out of genes) and biosensing technology [6,8].

Thus, this review is focused on providing an extensive knowledge of TDs in terms of their pathology and epidemiology, with more focus on mosquito-causing diseases. Consequently, the review provides background knowledge on the mechanism of CRISPR/Cas systems in prokaryotes, classification of Cas systems, and the application of CRISPR as a gene editing tool. The application of CRISPR/Cas systems has been shown to aid in disease diagnosis and treatment and the generation of Genetically Modified Organisms (GMO) in both plants and animals. Therefore, this review discusses the use of this technology in biosensing and disease detection.

The remaining part of the review is as follows: Section 2 overviews the concepts of TDs and classification of TDs based on pathogens (which includes bacteria, viruses, and parasites). Section 3 presents the discovery of CRISPR/Cas systems in prokaryotes (based on adaptation, expression, and interference) and the classification of Cas systems. Section 4 overviews the concept of the CRISPR/Cas gene editing tool. Section 5 provides the up-to-date literature on the application of CRISPR/Cas-based biosensors for the detection of TDs. Section 6 presents the open research issue and provides conclusions.

2. Tropical Diseases (TDs)

The study and classification of TDs became a hot topic during the era of exploration and colonialism by the British, American, Portuguese, Spanish, French, etc., who came in contact with these type of diseases in tropical regions. The study, diagnosis, and treatment of these diseases led to the establishment of tropical medicine. Increased research in this field has led scientists to understand the mode of transmission, vectors, and symptoms of these diseases during the 19th century. Moreover, pathogens such as viruses, bacteria, fungi, and parasites that are associated with TDs were identified, as well as vectors such as lice, mosquitoes, fleas, etc., and other TDs associated with food and water contamination [2,4].

Many TDs spread as a result of interactions and complex cycles of transmission between human primates and animals such as invertebrates (e.g., flies, mosquitoes, snails) and vertebrates (e.g., livestock, dogs, cats, bats, snakes, etc.). The widespread and reemergence of TDs depends on several factors including an increase in population or population growth, global warming, exploration, migration, deforestation, meteorological events such as flooding, urbanization, etc. However, a change in environmental conditions such as extreme weather conditions and variations in rainfall, temperature, and humidity have influenced the widespreadness of TDs compared to other factors. Variations of rainfall and temperature have both been associated with influencing pathogen and vector replication and reproduction, as well as vector metabolism, host distribution, and the selection of habitats for breeding [9].

2.1. Transmission of Tropical Diseases

In both tropical and temperate climate regions, numerous viral and bacterial diseases are spread via several routes including transmission from one person to another through coughing, sneezing (airborne disease), or sexual contact (sexually transmitted diseases). Example of airborne diseases include tuberculosis, measles, and respiratory syncytial virus. TDs can be also be transmitted through drinking contaminated water and food sources (also known as waterborne and foodborne diseases, respectively). The mechanism of transmission of the majority of these diseases depends on an intermediate carrier also known as a vector. These organisms or carriers harbor these pathogens from an infected person or animal (zoonotic) and transfer it to others. Most often, these pathogens undergo mutation or developmental changes within the carriers which make them more virulent and difficult for the human immune system to fight [9,10].

2.2. Classification of TDs

There are several ways in which TDs can be classified. However, the most common classifications are based on the type of pathogen (such as viruses, bacteria, parasites, etc.), vectors, or carriers (such as ticks, mosquitoes, flies, etc.), which are also termed as arthropod-borne diseases and also based on concern (neglected and non-neglected TDs) [2,4]. When these diseases are transmitted by arthropods (such as flies, ticks, or mosquitoes), they are termed arboviruses or arthropod-borne viruses [2].

2.2.1. Viruses

Virus-causing diseases are one of the most common and widely distributed pathogenic diseases in nature. Unlike bacteria that store their genetic content in the form of DNA, viruses store their genetic constituent in the form of RNA. They invade and hijack the host's nucleic acid replication system which provides the necessary machinery to replicate new viral particles [11].

Dengue Virus (DENV)

Dengue fever is a disease caused by positive-stranded RNA containing the virus known as dengue virus, which is transmitted by a mosquito-borne flavivirus. *Aedes aegypti* mosquitoes are regarded as the main carriers of dengue virus and can transfer this virus during feeding on human primates (also known as Human-to-Mosquito Transmission). Moreover, medical experts also report the possibility of maternal transmission (from pregnant mothers to their babies). The most common acute symptoms of dengue fever include severe pain in the muscles, joints, ocular inflammation, headache, nausea, vomiting, rashes, swollen glands, etc. When this virus infects infants and children, it causes "dengue hemorrhagic fever" which leads to critical conditions such as shock (also known as "Dengue shock syndrome") and circulatory system failure. Despite the prevalence of DENV around the world, there is no specific medication against the virus. However, doctors control the disease using medications to lower fever, relieve pain, prevent dehydration, manage bleeding, etc. [12].

In terms of epidemiology, DENV is found in many tropical and subtropical areas around the world and has been reported in more than 100 countries including Africa (Ivory Coast, Seychelles, Reunion Island, Cape Verde, etc.), Asia (Bangladesh, Afghanistan, China, Cambodia, Indonesia, India, Pakistan, Malaysia, etc.), America (Brazil, Peru, Ecuador, Nicaragua), and Australia. DENV has caused several endemics and epidemics around the world, with the incidence of the virus having recently increased due to human factors such as deforestation, massive urbanization, and global warming, which has expanded the regions inhabited by the Aedes mosquito vector. Approximately 400 million cases and more than 20 thousand deaths are reported almost every year, with more than 3 billion people at risk. The first outbreak of DENV dates back to 1779 in Indonesia and Egypt [13]. The disease was also recorded in North America in 1780 and it has reemerged over the years. In 2010, more than 1.5 million cases of DENV were reported in both South and North America. However, the largest number of infected cases was reported in 2016 in the United State of America (USA), with more than 2.38 million cases [14]. Several countries continue to report an increased number of cases daily, with Brazil having the highest number with more than 167 thousand as of March 2022.

Zika Virus (ZIKV)

ZIKV is another mosquito-borne disease that is predominant in several tropical and subtropical areas of West Africa, East Africa, South America, and Asia. The virus is a single-stranded positive-sense RNA virus that belongs to the "*Flaviviridae*" family. ZIKV shares numerous characteristics with other flaviviruses such as DENV, yellow fever virus, West Nile virus, and Japanese encephalitis. Aedes mosquitoes are regarded as the main carriers of ZIKV and can transfer this virus during feeding on human primates. During feeding, the virus is injected by mosquitoes which further replicates in dendric cells and is subsequently transported in the blood to other organs and tissues [15,16].

The virus can be acquired in the laboratory, through sexual intercourse or blood transfusions, or via the exchange of other bodily fluids such as breast milk, saliva, or the urine of an infected patient. The vector-borne transmission of the virus occurs in two cycles known as the sylvatic and urban cycles. The sylvatic cycle revolves around transmission of the virus by arboreal mosquitoes to non-human primates (NHPs), while the urban cycle revolves around transmission between human primates and urban mosquitoes. Scientists have also identified the virus antibodies in animal species such as goats, sheep, buffalo, lions, elephants, zebra, hippos, etc. The virus has been associated with Guillain-Barre syndrome in adults and microcephaly, arthrogryposis, ophthalmological defects, hearing defects, and cerebral malformations in children. The majority of infections caused by ZIKV are asymptomatic [16,17].

In terms of epidemiology, the virus was first isolated from the sentinel rhesus monkey in Uganda in 1947, while the first human isolation of the virus was reported in Nigeria in 1952. Since then, the disease has caused several epidemics and endemics around the world. Despite ZIKV having caused several health burdens, it was not until 2016 that the WHO declared it as a global health emergency due to the outbreak of the disease in South America. Just like many pathogenic tropical viruses, there is no specific antiviral drug or vaccine against ZIKV. Almost 100 thousand cases were reported in 2016, 609 in 2017, 1800 in 2018, and 15 cases in 2019 using EpiWATCH [18].

Yellow Fever Virus

Yellow fever virus is from the *Flaviviridae* family which causes yellow fever. It is related to *Japanese encephalitis*, St. Louis encephalitis, and West Nile virus. Yellow fever viruses are transmitted to people through a carrier known as Haemagogus or *Aedes mosquitoes*. These mosquitoes acquire the virus through feeding on infected animals or humans and transmit it to other primates. Thus, people infected by yellow fever virus through *Aedes mosquitoes* are referred to as being "viremic". Yellow fever has three transmission cycles which include sylvatic (jungle), where the virus is transmitted by mosquitoes from NHPs, such as monkeys to humans visiting the jungle; savannah (intermediate), where the virus is transmitted from mosquitoes directly to humans (human to human); or from NHPs to humans. Urban transmission is initiated by viremic humans who have visited the savannah or jungle region and urban mosquitoes which feed on the infected person and transmit the disease to other humans [19,20].

Some of the mild symptoms associated with this disease include headache, fever, chills, back pains, weakness, fatigue, vomiting, and nausea. When left untreated, it can lead to critical or severe conditions such as liver, kidney, and heart failure or malfunctions, shock, jaundice (yellow skin), bleeding, etc. The mortality rate is high, as more than 50% of people infected with the virus die of the disease. There is no specific drug against yellow fever diseases; however, physicians prescribe medications that relieve pain, fever, and aches [21].

Unlike DENV that is prevalent in almost every continent, yellow fever is limited to Africa where it originated and has caused several epidemics in South Africa and other African countries such as the Democratic Republic of Congo (DRC) and Angola [20]. Several incidences of the disease have also been reported in Latin America, with more than 12 South American countries affected. Despite the fact that there are vaccines against the disease, the prevalence of the disease continues to spread, resulting in more than 70 thousand deaths per year. The increase incidence of the disease is associated with the widespread distribution of *Aedes mosquitoes* as a result of climate change [22]. Yellow fever has been recognized as a disease of significant public concern due to it pathology and high mortality rate in both human and NHPs. However, little is known about why the cases cease in some years and appear in other years, and what promotes the strong seasonal trends [21].

Rotavirus

Rotavirus is a pathogenic double-stranded RNA virus from the *Reoviridae* family. The name "Rota" is derived from the Latin word meaning "wheel". Rotavirus is regarded as one of the most common pathogens that are detrimental in terms of pathology and is associated with causing watery diarrhea in children under 5 years of age. Other symptoms of the rotavirus-causing disease apart from severe dehydration include fever, nausea, vomiting, etc. Human primates are the reservoir of the disease which is found in the gastrointestinal tract and stool. The disease can be transmitted from person to person through fecal–oral routes (i.e., injecting infected food or water) and fomites (environmental surfaces contaminated by the stool of infected patients). The incidence of rotavirus is also reported in NHPs such as mammals, e.g., pigs [23,24].

The history of rotavirus dates back to the 1970s when several pediatricians and other medical experts embarked on studies to explore the causes of diarrhea in children as a result of striking mortality rates ranging from 3–12 million per year. Regions of high incidence include Bangladesh, Peru, and Guatemala which recorded more episodes of diarrhea cases. As a result of research using instruments such as electron microscopy and other biomedical instrumentation, scientists discovered several viral-causing diarrheas such as Norwalk agent and rotavirus (which appears to be a wheel-shaped virus), as well as other causing pathogens including different species of bacteria and parasites [23,24].

Globally, more than 2.7 million incidences are recorded and 600 thousand children die as a result of the virus annually, with the majority of deaths reported in India (i.e., with more than 100 thousand deaths yearly). Moreover, the majority of cases are reported in sub-Saharan Africa and South-East Asian countries. It is estimated that the disease leads to more than 500 thousand deaths in most of the underdeveloped countries. In the US, rotavirus is regarded as the most common cause of severe gastroenteritis in children [23,25].

Scientists over the years have developed vaccines against the virus, with the first developed in the USA in 1998 known as Roatshield, which was later withdrawn due to rare adverse effects. Currently, there are several vaccines used to treat rotavirus (such as Rotavac, Rotateq, Rotasil, Rotarix, etc.), in more than 100 countries around the world. The increase in vaccination has led to a decrease in incidence, hospitalization, and mortality among infants. Despite the massive level of vaccination, the disease continues to cause concern in many underdeveloped countries with low access to vaccines and quality medical care [24].

Human Immunodeficiency Virus

HIV/AIDS is one of the most common diseases that affects the immune system. It is caused by human immunodeficiency virus which belongs to the genus of viruses known as *"lentiviruses"*. In terms of its pathophysiology, HIV has been described to overcome or overpower the immune system's T-cells known as CD4 helper cells, rendering the immune system susceptible to invasion from other pathogenic agents and cancers. As a result of the decline in response to foreign invaders by the immune system, HIV is accompanied by the term AIDS (Acquired Immune Deficiency Syndrome). The virus can be transmitted in several ways as a result of the exchange of bodily fluid such as blood, breast milk, vaginal secretions, and semen. The virus can also be transmitted from an infected pregnant mother to her baby. Some of the mild symptoms of the disease include fever, headache, sore throat, rashes, diarrhea, cough, swollen lymph glands, weight loss, etc. [26,27].

In terms of epidemiology, HIV is believed to originate from West Africa where it was transmitted to humans by a subspecies of chimpanzee. The disease is among the list of the most critical diseases that have emerged in the history of humanity. By 1996, the disease had already infected more than 13 million people within sub-Saharan Africa. The disease was declared epidemic in 1989 by the WHO as a result of an increase in the number of cases [28]. Despite major advances in diagnosis and treatment of HIV over the past two decades, it still remains a global concern. Currently, there is no specific drug against the virus. Even though there has been a decrease in the number of deaths cause by the virus, it is still prevalent in poor countries with substandard medical care systems. HIV has spread to almost every country and it was estimated that more than 37 million people were living with the virus in 2020 [29].

Ebola Virus

Ebola virus is one of the reemerging diseases causing severe health issues in African countries. It is formerly known as Ebola hemorrhagic fever. Symptoms of the disease cause by the virus include fever, hemorrhage, headache, and vomiting diarrhea. The mode of transmission of the disease is still unclear but medical experts believe the virus can be acquired as a result of direct contact with bodily fluid such as bloods and other secretions from infected patients as well as contact with surfaces contaminated with the virus. Scientists have categorized it as zoonosis and linked the disease with fruit bat and porcupines [30,31].

Ebola virus originated from two African countries, including the DRC and Congo, in 1976. The virus reemerged in West African countries, with early incidences in Liberia, Guinea, Sierra Leone, and Nigeria in 2014, and is regarded as the most serious health emergency crisis in the region. As of 2015, there had been more than 28 thousand reported cases and more than 11 thousand deaths. The average fatality rate of the disease is 50%; however, fatality rates in the past outbreak have varied between 25–90%. Even with advances in the diagnosis of the disease using advance technology, the outbreak of Ebola still remains intermittent and unpredictable. There have been more than 30 outbreaks of Ebola since 1976 [30]. The two recent outbreaks were reported in the DRC in 2018, with more than 3 thousand cases, and on 7 February 2021 [31,32].

2.2.2. Bacteria

Bacteria are among the most abundant and ubiquitous microbes in nature. Bacteria can be classified as either pathogenic or non-pathogenic (e.g., microbiomes). They can also be classified as Gram positive or Gram negative. Some of the pathogenic bacteria that cause diseases include *Mycobacterium tuberculosis*, *Vibrio cholerae*, *Escherichia coli*, etc. [33].

Tuberculosis (TB)

TB is one of the most common bacterial diseases caused by bacteria know as *Mycobacterium tuberculosis*. It was discovered in 1882 and its mode of transmission was first reported in 1909. It is a slender, rod-shaped microbe with length ranging from 1–10 mm and strict aerobes (i.e., needing oxygen to survive). Tuberculosis is an airborne disease that is transmitted from an infected patient to others via sneezing, coughing, talking, etc. Depending on the environment, the bacterial particles can remain suspended in the air for hours and thus can be transmitted as a result of coming in contact with surfaces contaminated with the bacilli [34].

The pathogenesis of the bacteria occurs in the lung's alveoli where it causes pulmonary tuberculosis. A few weeks after exposure, a granuloma is formed as a result of the immune system response against the bacilli. When the bacteria spread to other tissues in the body it is termed as "systemic miliary tuberculosis". Treatment of tuberculosis depends on the severity of the disease. Pulmonary tuberculosis is mostly treated using antibiotics. However, *Mycobacterium tuberculosis* is becoming resistant to drugs and thus has increased virulence [35].

TB still remain a global health issue despite the use of antibacterial drugs against the bacteria. As of 2015, there are more than 10 million people suffering from the disease, with 10% mainly children and 12% people suffering with HIV/AIDS. As of 2015, the number of deaths associated with TB was estimated to be around 2 million. In the last few years, cases of TB have declined marginally. Even though the disease has been controlled in many countries, it is still a health issue in many underdeveloped countries and continues to threaten to become an increasing burden due to both extensive drug resistance and multi-drug resistance [35].

Cholera

Cholera is another bacterial disease that is associated with diarrhea. It is caused by a Gram-negative bacterium with a coma-like shape known as *Vibrio cholerae*. Cholera can be transmitted through the fecal–oral route as a result of eating food or drinking water contaminated with the bacteria. When the bacterium enters a host cell, it secretes toxins which leads to symptoms such as diarrhea, vomiting, abdominal pain, and hypovolemic shock. Factors that increase the risk of the disease include lack of access to clean and sanitized water, people with O blood group, living in overcrowded societies, use of antihistamine and proton pump inhibitors, etc. [36,37].

Despite progress in research regarding the diagnosis and treatments of diseases, cholera continues to be a burden in many countries. In terms of outbreaks, scientists have identified two serotypes known as O1 and O139 which causes disease, while more than 190 serotypes are non-pathogenic. O1 has been associated with the most recent outbreaks in Bangladesh and Kerala, India. O139, on the other hand, has caused sporadic outbreaks in some regions within Asia [37,38]. Despite the fact that the majority of positive cases and deaths toll are underreported, it was estimated that there are more than 4 million cases of cholera yearly and more than 140 thousand deaths globally. The disease is found to be endemic in many countries within Asia and Africa, while cases have been reported in countries within the Caribbean, Middle East, and South and North America [38,39].

Escherichia coli

E. coli is another bacterium that causes bacteremia in some developed countries. The bacteria are found in the lower intestine of blooded animals. There have been several strains of *E. coli* identified and the majority do not cause infection, while a few are pathogenic and have been found to cause food poisoning and diarrhea. An example of this bacteria is the Shiga toxin-producing *E. Coli* O157:H7. This strain is an enterohemorrhagic type that causes diarrhea hemolytic-uremic syndrome and hemorrhagic colitis in humans. It is classified as both a food and water-borne disease that is transmitted via the fecal–oral route as a result of consuming uncooked meat and contaminated liquid, including raw meat and vegetables [40]. Pathogenic *E. coli* are known to cause travelers' diarrhea and kidney problems. Common symptoms include abdominal cramp, vomiting, fever, and diarrhea. Outbreaks of this pathogenic bacteria have been reported in Japan, the USA, and Scotland [41].

2.2.3. Parasites

Parasites are group of organisms that live in or on another organism known as the host, at whose expense they obtain their nourishment while simultaneously infecting the host. Parasites can be classified as single-cell organisms (e.g., protists) and multi-cell organisms (e.g., helminths or worms). Parasites ranges from micro-size to macro-size organisms. Even though some parasites can be found intracellular, the majority live extracellular inside the host and are mostly found in the gut, blood, lymphatics, etc. Unlike bacteria and viruses that replicate inside host primates, parasites undergo complex developmental transformation within the host and can reproduce sexually and asexually. Examples of parasitic diseases and pathogens include *Plasmodium* (malaria), *Trypanosoma* (African sleeping sickness, Chagas disease), hookworm (ancylostomiasis), roundworms (Ascaris), tapeworm (Dipylidium caninum disease), *Leishmania* (Leishmaniasis or kala-azar), etc. [42].

Malaria

Malaria is the most common parasitic disease globally. It is cause by protists known as *Plasmodium*. There are different species of *Plasmodium*; however, *Plasmodium falciparum* is identified as the most virulent and the leading cause of death among *Plasmodium* species. *Plasmodium* is transmitted to human primates by anopheline mosquitoes during feeding. After injection of the parasites by the carrier, the infection cycle begins in the liver cells followed by the red blood cells where the parasites consume hemoglobin. The completion of the cycle ends in the erythrocytes, where the parasites divide and infect more red blood cells. The general symptoms of malaria include fever, weakness and fatigue, sweating, nausea, vomiting, diarrhea, headache, abdominal pain, etc. [43,44].

Malaria is more prevalent in sub-Saharan Africa countries, where there are widespread untreated water bodies. Malaria cases are estimated to be around 300 million with more than 1 million deaths every year. Malaria has also reemerged in countries that had been declared malaria-free. Outbreaks of malaria continue to be an issue in Africa, the Amazon region, and Asia. The WHO and other international organizations have launched several campaigns and projects to eradicate malaria globally but it still remains elusive. Factors associated with persistent malaria include resistance to insecticide and drugs, lack of adequate healthcare facilities and sanitation programs in underdeveloped countries, and lack of priority concern from international bodies [45,46].

Trypanosomiasis

Trypanosomiasis is another disease that is cause by species of the genus Trypanosoma. An example of trypanosomiasis is Chagas disease, also known as American trypanosomiasis that is cause by *Trypanosoma cruzi*. Unlike malaria, trypanosomiasis is transmitted by bugs which feed on infected feces and enter into human primates through the mouth, nose, skin, eyes, etc. Symptoms of this disease include inflammation or swelling of the lymph nodes and fever, while in critical conditions it can lead to cardiac malfunction, digestive disorders, and death [47].

Another common disease-causing trypanosomiasis is known as sleeping sickness or African trypanosomiasis, which is transmitted by tsetse flies (Glossina species). Species associated with this disease include *Trypanosoma brucei rhodesiense*, which is found in South and Eastern Africa, and *Trypanosoma brucei gambiense*, which is most common in Central Africa and West Africa. The common acute symptoms of African trypanosomiasis include weakness, dizziness, fever, and headache. In a critical condition, the disease can cause neurological disorders with symptoms such as delusions, hallucination, seizures, etc. [48,49]. Epidemics of sleeping sickness have caused concern in the past. However, as a result of intervention by both national and international organizations, the disease is well-controlled with cases of less than 600 reported in the DRC in 2020 [50].

Leishmaniasis

Leishmaniasis is a parasitic disease that is cause by protists known as the Leishmania genus. Scientists have identified about 20 species that causes disease to human primates and NHPs (i.e., mammals). An example of Leishmaniasis is Cutaneous Leishmaniasis, which is locally known as oriental sore, Delhi boil, or Baghdad ulcer. Another example is Visceral Leishmaniasis, which is locally known as kala-azar in India and refers to black sickness. In recent years, the number of cases has surged from less than a million to 1.2 million. Visceral Leishmaniasis is regarded as the most dangerous form of Leishmaniasis. Leishmania can be transmitted as a result of a bite from infected phlebotomine sandflies. This species is found within the macrophages and plays a crucial role in fighting against invading microorganisms in the host's body [51]. The symptoms of Leishmaniasis include weight loss, increase pigmentation, fever, and swelling of the liver and spleen. Apart from humans, other reservoir hosts identified include dogs and rodents [51,52].

In terms of epidemiology, the disease is prevalent in both tropical and subtropical regions, with recent cases in Sudan and India. Few cases have been reported in the US and Southern Europe, while Australia and Antarctica are the only continents with no reported cases. According to the WHO, there are more than 10 million recorded cases with 300 million people at risk in more than 90 countries. The disease is predominant in rural settlements but can also be found in the outskirts of cities [53].

2.3. Neglected Tropical Disease (NTDs)

The last century has witnessed a decline in the incidence of and elimination of numerous TDs in the majority of developed countries. However, millions of people are still affected by these types of diseases, especially in underdeveloped countries which contribute to high mortality rates. These types of diseases are termed as NTDs. Example of NTDs include leprosy, Guinea worm disease, African sleeping sickness, rabies, Leishmaniasis, Schistosomiasis, Fascioliasis, dengue, Dracunculiasis, Onchocerciasis, Chagas disease, yaws, hookworm, trachoma, etc. [54].

The prevention and control of NTDs in underdeveloped countries are highly challenged by several factors such as the socioeconomic status of the regions, a lack of medical equipment, a lack of adequate response and concern from international organizations, a lack of awareness, etc. NTDs can be found in some regions located within Africa, Asia, and Latin America. The majority of NTDs are associated with rural areas and regions which lack access to hygienic food, clean water, and safe ways of waste disposal [32].

The primary ways of controlling or preventing NTDs include controlling the vectors or via massive drug administration. As carriers of pathogens, vectors play crucial role in disease pathways. Thus, controlling vectors such as black flies and mosquitoes that transmit disease as well as improving environmental hygiene and water sanitation are highly crucial for controlling and preventing NTDs. Consequently, massive drug administration is another effective way or intervention in eliminating NTDs. Diseases that can be eliminated using this intervention approach include trachoma, Onchocerciasis, Dracunculiasis, Schistosomiasis, lymphatic filariasis, and soil-transmitted helminths (hookworm or Ascaris) [55].

3. CRISPR in Prokaryotes

The CRISPR systems along with Cas proteins are highly diverse adaptive immune mechanisms used by many bacteria and archaea to protect themselves from attacks by viruses, plasmids, and other foreign nucleic acids. CRISPR consists of short, highly conserved repetitive sequences (23–44 bp long) separated by spacers. These spacers are unique sequences and are usually derived from phages or the plasmid's DNA. This adaptive system can learn to recognize and cut specific NA regions of invading pathogens and store them [56].

CRISPR evolved as an immune response or mechanism against phages. The basic mechanism of the CRISPR/Cas system arises from the need to obtain viral DNA or RNA, with most archaea (~87%) and bacteria (~47%) being clustered in normal short intervals. Intact CRISPR loci within the genome include a series of CRISPR arrays, CRISPR-related protein (Cas) genes, and direct repeats separated by multiple spacers. With the onset of the virus, the CRISPR/Cas locus triggers a three-step immune response, "adaptation–expression–interference", which destroys phages that invade the host cell [6].

Technically, the CRISPR/Cas system has only two components: (I) Cas protein, DNA, or RNA cleavage protein that promotes adaptive immunity in the process of adaptation, expression, and interference in prokaryotic cells, and (II) for cleavage of a target nucleic acid (DNA or RNA) as a known RNA molecular guide RNA which is programmed to navigate the system to recognize, bind, and cleave target NA [57]. The adaptation stage occurs when bacterial or archaeal cells first come in contact with viral DNA. The CRISPR loci translate Cas genes into Cas proteins (Cas9, Cas2, and Cas1). These Cas proteins survey for the viral DNA, cut part of it (known as spacer), and store it in the CRISPR array's leader strand. The second stage, known as the expression stage, is only initiated when the viral DNA attack again. The bacterial or archaeal cell's CRISPR array transcribes its stored spacers into small non-coding RNA, known as pre-CRISPR RNA, which link with TracrRNA through base pairing and form hybrid RNA or matured CRISPR RNA. The CRISPR RNA is employed in the third stage, known as the interference stage, where it forms a complex with an effector Cas enzyme (such as Cas9 which is translated from the Cas genes adjacent in the CRISPR loci). This complex locates the viral DNA as a result of the unique Protospacer Adjacent Motif (PAM) sequence and destroys the viral DNA, leading to complete immunity, as shown in Figure 1 [58–60].

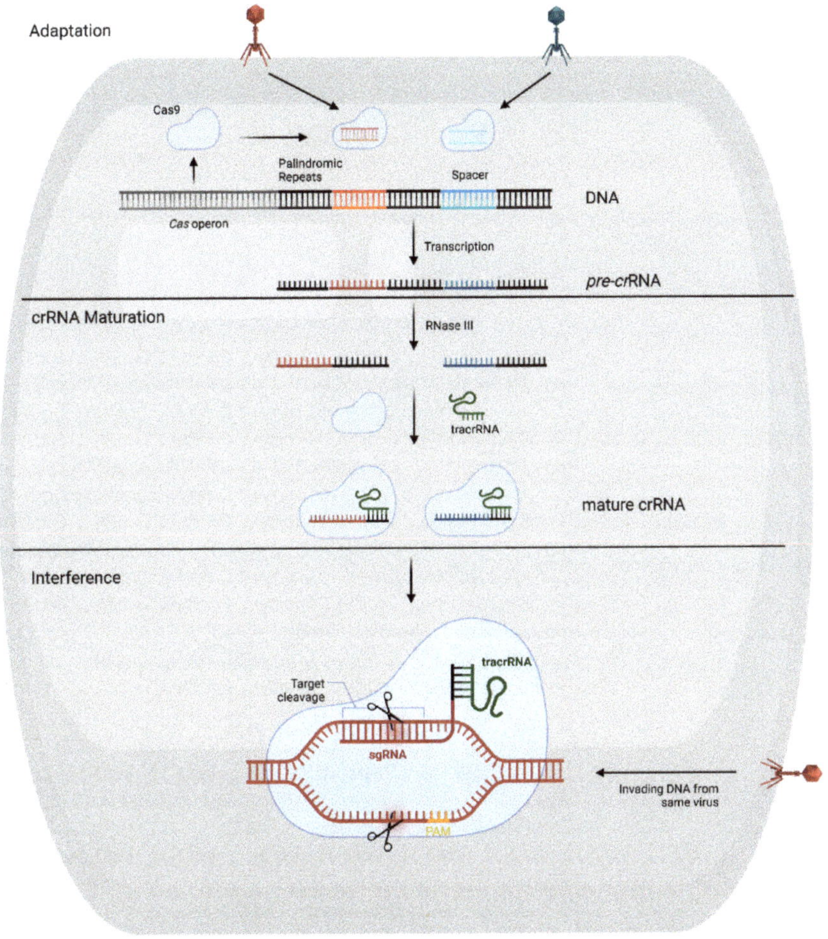

Figure 1. Schematic diagram of the CRISPR adaptive immune system against bacteriophage. (Adapted from Ref. [61] with free permission. Created with BioRender.com accessed on 20 September 2022).

3.1. Classification of Cas Systems

The CRISPR-Cas system can be divided into two classes depending on the number of effector Cas used in the interference stage. Class I contains multiple Cas effector complexes. Class II requires only one Cas protein. Based on their properties, the CRISPR/Cas system class can be divided into several different types, which are further subdivided into subtypes corresponding to specific Cas proteins. Recent studies have shown that types I, III, and IV belong to class I, and types II, V, and VI belong to class II. Today, the CRISPR-Cas system offers new methods of biosensing with its ability to identify single-base mismatches in target nucleic acids. Many Cas effectors possess specific (cis-cleavage) and non-specific (trans-cleavage) nucleolytic activities. Type II Cas9, Type V Cas12a, Type VI Cas13a, and Type V Cas14 are widely used along with guide RNA (gRNA) complexes to target-specific DNA/RNA [62]. The classification of CRISPR/Cas systems is presented in Table 1.

Table 1. Classification of CRISPR/Cas systems.

Class	Type	Adaptation	Pre-CrRNA Processing	Effector Module	Target Cleavage
Class 1	I	Cas1, Cas2, and Cas4	Cas6	Cas7 and Cas5	Cas3
	III	Cas1 and Cas2	Cas6	Cas7 and Cas5	Cas10
	IV	-	-	Cas7 and Cas5	-
Class 2	II	Cas1, Cas2, and Cas4	RNaseIII	Cas9	Cas9
	V	Cas1, Cas2, and Cas4	-	Cpf1 (Cas12) and Cas14	Cpf1 (Cas12) and Cas14
	VI			Cas13	Cas13

3.1.1. Cas9

Cas9 is a double-spin RNA-driven type II DNA cleavage protein. Only double-stranded DNA (dsDNA) is required as the catalytic substrate. An adjacent protospacer motif (PAM) is required for the target DNA. Unlike other Class II effector types, Cas9 uses Rnase III to process the transactivation precursor RNA (tracrRNA) and CRISPR RNA (crRNA) complex before binding to dsDNA. Mature gRNA begins with a nucleotide spacer (nt) 20–24, followed by a tracrRNA: crRNA double chain. gRNA is also chimeric and can form single-stranded guide RNA (sgRNA). Cas9, which has multiple domains, interacts with mature crRNA to stabilize the crRNA and change its conformation, facilitating the binding of the next target. In the presence of dsDNA, Cas9 first looks for the PAM sequence, then recognizes the seed region and forms Watson Crick base pairing between the target dsDNA and the spacer. The RuvC and HNH domains cleave target strands (TS) and non-target strands (NTS) (3 nt upstream of PAM) to introduce blunt-ended double-strand breaks (DSBs). As mentioned earlier, the cleaved product still binds to the Cas protein and is released very slowly. In addition, Cas9 catalytic deficiency (dCas9) is acquired by mutations in the nuclease domain, which retains only its DNA-binding ability [63].

3.1.2. Cas12a/Cas12b

Cas12a/Cas12b produces mature crRNA and directs the Cas protein to bind to the target DNA. The length of the mature crRNA was 42×10^{44} nt. It begins with a 19 NT direct iteration sequence in which the 19th U base is strictly retained, followed by a 23–25 nt spacer. The interaction of Cas12 crRNA triggers the conformation switch of Cas12a, exposing the active site of RuvC. When dsDNA containing the PAM-rich T sequence at the 30-end is perfectly fitted to the crRNA spacer, it forms an R-loop with crRNA. NTS is placed in the active site of RuvC for the next cleavage. After cleavage, Cas12a allows the release of the truncated product, revealing the active site of RuvC. This produces a staggered DSB with a 5 nt overhang at the 50th edge. Interestingly, Cas12a's transactivity allows it to cleave adjacent ssDNA without the need for a specific sequence. In contrast to Cas12b, dsDNA-triggered Cas12a has higher trans-cleaving efficiencies than ssDNA-triggered ones [61,62].

3.1.3. Cas13a/Cas13b

Compared to other Class II Cas proteins, Cas13 showed cis and trans-cleaving activity against single-strand RNAs (ssRNAs). Cas13 interacts with (pre) crRNA to cause conformational change when it recognizes RNA with a 3′-protospacer flanking sequence (PFS, A/U/C). When the seed region closely matches the target from the center of the spacer, it then extends throughout the spacer to form a double chain. In addition, the HEPN domain approaches the construction of complex active sites on the outer surface of Cas13, causing Cas13 to function as a non-specific Rnase. Studies have shown that a 20-base pair (bp) guide target double-stranded RNA is essential for activation of the catalytic site of the HEPN domain. Furthermore, Cas13 lacks a specific cleavage site but shows a cleavage preference for U [62,64].

3.1.4. Cas14

Cas14 is one of the most recent characterized Cas systems which shares similar traits with CRISPR/Cas type V. As a new member of the CRISPR effector, Cas14 is much smaller in size (40–70 kDa) compared to other Cas proteins in the Class II system and typically has a molecular size of 100–200 kDa. Unlike Cas9, Cas14 is regulated by tracrRNA (crRNA double chain or sgRNA). Unlike Cas13 that cleaves RNA, Cas14 cleaves single-stranded DNA targets similar to Cas12. Cas14 can recognize foreign DNA without the need for PAM sequences. In addition, Cas14 cleaves the target ssDNA beyond the spacer protospacer double chain region, and its collateral cleavage efficiency increases with ssDNA elongation. Cas14 also exhibits collateral cleavage activity against DNA, which makes it vital for the direct detection of pathogenic bacteria as well as RNA after reverse transcription. Despite the fact that Cas14 shares a lot of similarities with Cas12, Cas14 has a lower on-target as a result of sensitivity of the internal-seed sequence to nucleotide mismatch, as well as a lower tolerance to nucleotide mismatch that lies between the target template and sgRNA [65]. The differences between Cas effectors are shown in Table 2.

Table 2. The difference between Cas effectors.

Differences	Cas9	Cas12	Cas13	Cas14
Domains	RubC and HNH	RuvC	2 HEPN	RuvC
Target	DSDNA	SSDNA	RNA	SSDNA
Organism derived from	*Streptococcus pyogenes* *Streptococcus thermophilus* *Staphylococcus aureus*	*Prevotella* sp. *Francisella* sp. *Lachnospiraceae bacterium* ND 2006 *Acidaminococcus* sp.	*Prevotella* sp. *Leptotrichia wadei*	*Extremopile archaea*
Types of cuts	Blunt	Staggered	-	-
TracRNA	Present	Absent	Present	Present
PAM sequence	NGG	T-rich	PFS	Not required

4. CRISPR/Cas as Gene Editing Tool

CRISPR/Cas9 has been shown to function as an adaptive immune system against viruses and phage through DNA binding by CRISPR RNA (crRNA) and DNA damage by Cas9 nuclease in bacteria. In genome editing, CRISPR/Cas9 functions with the help of a single guide RNA (sgRNA) that recognizes a target sequence (protospacer) in the genome of the host organism via complementary base pairs. The Cas9 nuclease then specifically creates a double-strand break (DSB) in the region close to the PAM sequence (Protospacer Adjacent Motif). A major advance in this area is the discovery of sgRNA. It was originally used in combination with Cas9 and made in vitro cuts at various DNA sites [3].

Unlike in prokaryotic cells, the CRISPR/Cas complex acts as an antiviral system to identify the genetic information of alien species (DNA or RNA fragments injected into the cell) and stores and shares it highly selectively and specifically. However, in the case of biomimicking the prokaryotic CRISPR/Cas mechanism in living cells, the guide RNA strand is synthetically designed to bind consistently with the DNA or RNA sequence of an exotic species. After finding the correct sequence, the CRISPR/Cas complex cleaves DNA or RNA into one or two nuclease domains (depending on the type of Cas protein) and creates nicks to make them single- or double-stranded [60].

Once the DNA/RNA is cleaved, the cells initiate a repairing mechanism known as Non-homologous End-joining (NHEJ), which is a natural way that cells stick together through insertions or deletions of nucleotides (known as indels). However, this method is prone to mutation and can lead to gene dysfunction or deactivation. Scientists can use this window to introduce a desired homologous DNA template through a process known as Homologous Repair (HR) or Homologous Directed Repair (HDR), as shown in Figure 2 [66].

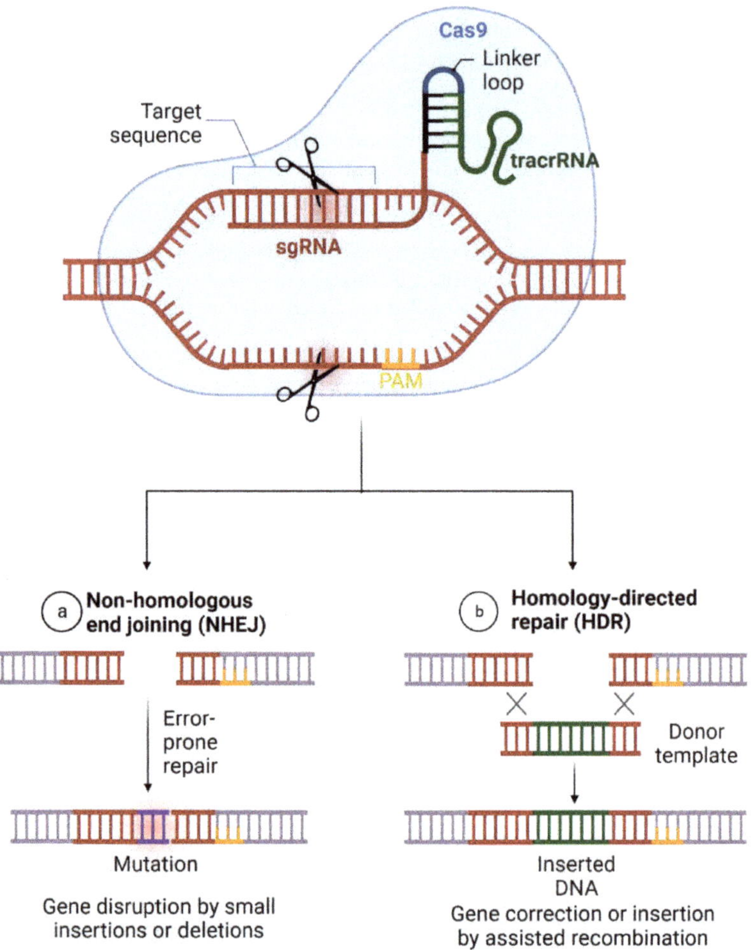

Figure 2. Gene Editing Using CRISPR/Cas system. (Adapted from Ref. [61] with free permission. Created with BioRender.com accessed on 20 September 2022).

CRISPR/Cas9 in Parasites

The discovery of CRISPR/Cas systems has opened the gateway to several applications related to gene editing of parasites. Applications include the use of CRISPR/Cas9-mediated gene drive to interfere with vector transmission of parasitic diseases, the choice of selectable markers, novel delivery and treatment approaches, understanding of the pathogenesis of parasitic organisms through gene manipulation, etc. Despite the prospects of CRISPR/Cas9 gene editing technology, it is hindered by several challenges which include off-targets, gene mutations, and complex morphology and the life cycle of these parasites. Thus, there is a need to develop novel approaches that will increase the efficiency of CRISPR/Cas9 gene editing technology and improve on-targets, enhancing gene mutation efficiency and overcoming issues involved in the host passage [67,68].

Since the discovery of CRISPR/Cas9, the system has been used in a wide variety of bioscience and biomedical studies to edit genomes of a wide range of model organisms (which include *Caenorhabditis elegans*, *Saccharomyces cerevisiae*, *Drosophila melanogaster*, etc.), generation of animal models, cancer treatment, stem cell research, somatic genome editing, correcting genetic diseases, neurobiology, and the treatment of infectious diseases [68].

There are a handful of studies that attempted to knock-in or knock-out genes from parasitic organisms. Among these studies is the one provided by Dong et al. (2018) [69]. The study, as part of an approach to control mosquitoes, targeted the agonist's journey of *Plasmodium* based on transmission-blocking using CRISPR/Cas9. The study proposed a CRISPR/Cas9 gene editing procedure which targeted a malaria vector known as *Anopheles gambiae* by inactivating fibrogen related protein 1 (FREP 1). The result of the study has shown profound suppression of malaria infection in adult mosquitoes (FREP1 knock-out mutants). Another study that focused on controlling mosquitoes was provided by Macias et al. (2020) [70]. The study proposed an embryo injection method based on Receptor-mediated Ovary Transduction of Cargo (ReMOTE), which is used to transport Cas9 ribonucleoprotein complex to the ovaries of an adult *Anopheles stephensi*. The outcome of the study demonstrated the efficiency of ReMOTE in delivering Cas9 and the subsequent development of heritable mutations in adult mosquitoes.

Unlike the study conducted by [69] which focused on mosquitoes, the study conducted by Zhang et al. (2017) [71] revolved around *Plasmodium*. The study targeted *Plasmodium yoelii* ApiAP2 genes which have been shown to play a significant role in parasite development. The study identified 24 genes and 12 were successfully knocked out using CRISPR/Cas9. However, evaluation of the gene knockout in the development of *Plasmodium* in both mice and humans have shown that some of the genes are critical for the development of *Plasmodium yoelii*.

Gene drive technology is gaining attention due to it scalable impacts on controlling infectious diseases. The use of CRISPR/Cas9 is becoming the most important machinery for the genetic manipulation of parasites and vectors [72,73]. An example of gene drive technology for disease control was proposed by Hammond et al. (2016) [74]. The study identified three genes which contribute to recessive female sterility in mosquitoes. The genes are introduced into each locus using CRISPR/Cas9. The evaluation of the impacts of the genes in controlling the population of mosquitoes using population modeling and cage experiments revealed that one of the genes met the minimum requirements for gene drive.

The study conducted by Burle-Caldes et al. (2018) [75] applied the CRISPR/Cas9 gene editing approach for rapid generation of *Trypanosoma cruzi* gene knockout mutants. The study focused on the disruption of the GP72 gene, which is achieved either through transfecting wild type *T. cruzi* with recombinant *Staphylococcus aureus* Cas9 bind with guide RNA or through transfecting *T. cruzi* stably expressing *Staphylococcus pyogenes* Cas9 along with SgRNA. Due to the absence of NHEJ repair in the parasites, the study showed that gene knockout in *T. cruzi* occurs through HDR instead of microhomology-mediated end joining (MMEJ). Moreover, disruption of these genes has resulted in abnormal morphology and few parasites had their flagellum detached from their body.

The first reported editing of the CRISPR/Cas9 gene in kinetoplastids, for *Leishmania donovani* [76] and Leishmania major [77], used a method of expressing Cas9 from an episomal plasmid. Target-specific SgRNA targets are in vitro transfected sgRNA transfection or plasmid transfection for in vivo transcription of sgRNA from RNA Pol I [76] or RNA Pol III [77] promoters. Donor DNA for directed repair results in precise modification [76]. Cas9-induced double-strand cleavage was repaired by a microhomology-mediated end joining (MMEJ) mechanism without adding donor DNA, resulting in a small deletion at the target site [76]. The use of CRISPR/Cas9 for the knockout of the lipophosphoglycan (LPG) gene in *Leishmania* spp. was proposed by Beneke et al. [78] with a CRISPR/Cas9 toolkit for rapid and precise gene modification by integration of donor DNA, using engineered cell lines and drug selection of mutants. All required sgRNA and donor DNA cassettes are generated by PCR in just a few hours without time-consuming DNA cloning.

5. CRISPR-Based Biosensor

The NA detection technique is one of the molecular diagnostic approaches that has been trending over the past few years. Apart from PCR and RT-PCR that are established as good standard approaches for the detection of viruses and other pathogens, other approaches such as NA hybridization and isothermal application techniques have been developed for clinical diagnostics. Despite the viability of these approaches, they are hindered by several challenges such as low specificity, low sensitivity, and the need for laborious procedure, chemical reagents, and well-trained medical lab technologists [7,66].

CRISPR-based biosensors are gaining interest from scientists around the world due to their sensitivity toward target NA. The need to develop simple, robust, sensitive, accurate, cheap, and POC diagnosis for clinical applications is growing every year. CRISPR toolbox is a Pandora's box with several Cas systems that allow scientists to delete and insert gene of interests. This remarkable feature allows scientists to develop CRISPR-based biosensors as a subsidiary of a NA-based biosensor that can either bind or cleave to target nucleic NA. Inside this Pandora's box are CRISPR/Cas9, CRISPR/Cas12, and CRISPR/Cas13 [79,80]. Recent studies have shown that Cas14 can be used as a promising tool for diagnosis and biosensing [62].

5.1. CRISPR/Cas9 or dCas9-Based Biosensors

The CRISPR/Cas9 is one of the most widely used Cas systems for gene editing. The process revolves around the use of a Cas effector along with single guide RNA which navigates through a matching target and cleaves it by inducing a double-strand break, as shown in Figure 3. The use of CRISPR/Cas9 for the detection of diseases can be classified into two approaches: (1) Cleavage-based biosensing (Figure 3A,B) and (2) Binding-based biosensing (Figure 3C,D). CRISPR/Cas9 has been harnessed along with other amplification techniques such as Nucleic Acid Sequence-based Amplification (NASBA)-CRISPR Cleavage (NASBACC) (as shown in Figure 3A) and CRISPR/Cas9 Triggered Isothermal Exponential Amplification (CAS-EXARP) (as shown in Figure 3B) reaction to form biosensing platforms for the detection of pathogens and discrimination between different strains. dCas9 is an inactive form of Cas9 which instead of cleaving the target, only binds to it. Scientists harnessed this feature to couple various modules such as split enzyme or fluorescent to develop a bioaffinity CRISPR-based biosensing platform [7].

Zhang et al. (2022) [81] developed a biosensing platform that use two pairs of dCas9 for the detection of *Mycobacterium tuberculosis*. The system was designed using pairs of dCas9 conjugated to the split halves of luciferase, termed as paired dCas9 (PC) reporter, as shown in Figure 3C. The Cas system was guided by two single guide RNA which activate luciferase. The binding of the complex generated highly intensified luminescent signals. Evaluation of the sensitivity of the biosensing system resulted in a sensitivity of 0.1 nM without DNA amplification and 35 aM with amplification using PCR (35 cycles).

The use of CRISPR-based optical Geno-biosensor for the detection of the Zika virus was developed by Pardee et al. (2016) [82]. The study employed CRISPR/Cas9 and isothermal RNA amplification which is able to discriminate between Zika genotypes with single-base resolution. The study also used DENV as a negative control. Evaluation of the optical Geno-biosensor has shown high specificity in discriminating between ZIKV and DENV. Moreover, the study conducted by Qui et al. (2018) [83] demonstrated the use of Rolling Circle Amplification (RCA) for isothermal amplification of the target (microRNA), with dCas9 effectors fused together with split Horseradish Peroxidase (HRP) protein to recognize and bind to the target in order to activate the colorimetric change of Tetramethylbenzidine (TMB), as shown in Figure 3D.

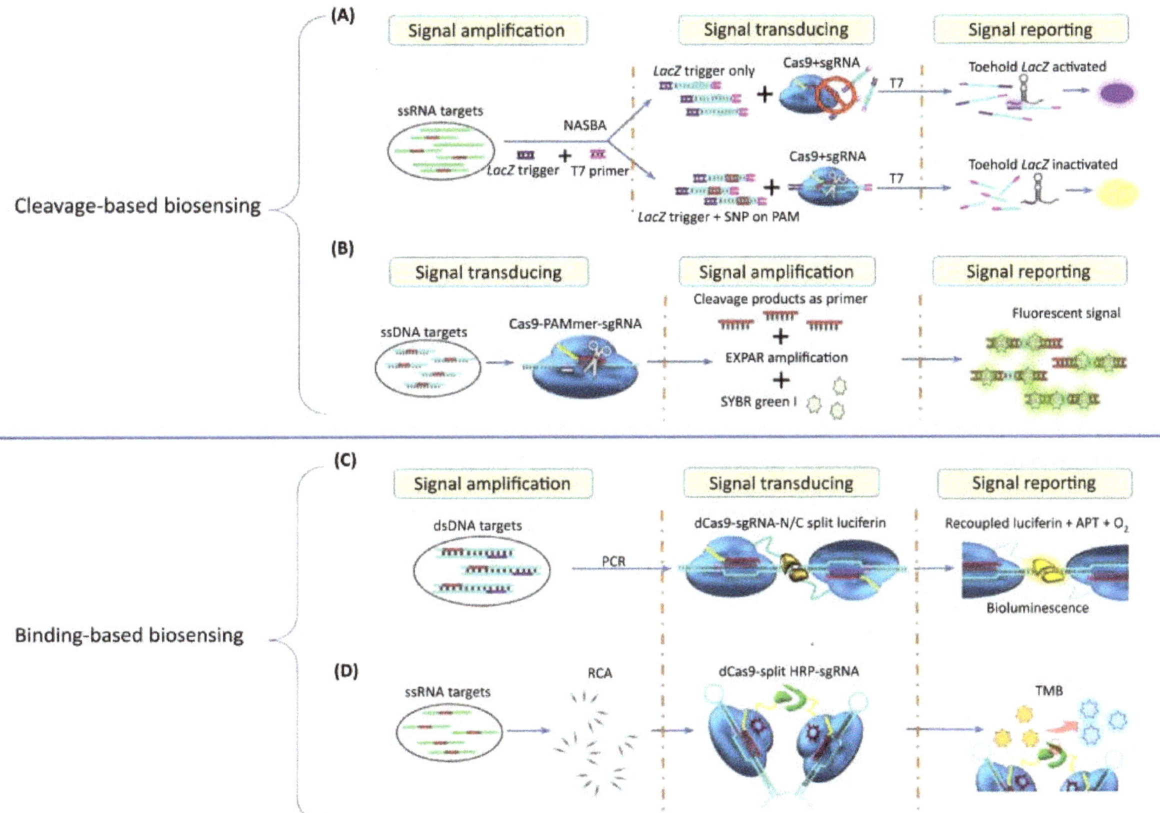

Figure 3. The use of CRISPR/Cas9 and dCas9 for the detection of diseases. (Reprinted with permission from Elsevier [79]. Copyright and Licensing.) The use of CRISPR/Cas9 for the detection of diseases can be classified into two approaches: (**A,B**): Cleavage-based biosensing. (**C,D**): Binding-based biosensing. (**A**). Nucleic Acid Sequence-based Amplification (NASBA)-CRISPR Cleavage (NAS-BACC). (**B**). CRISPR/Cas9 Triggered Isothermal Exponential Amplification (CAS-EXARP). (**C**). dCas9 conjugated to the split halves of luciferase, termed as paired dCas9 (PC) reporter. (**D**). Rolling Circle Amplification (RCA) for isothermal amplification of the target (microRNA), with dCas9 effectors fused together with split Horseradish Peroxidase (HRP).

5.2. CRISPR/Cas12-Based Biosensors

Unlike Cas9, which possesses both RuvC and HNH domain which induce double-strand break, Cas12 only possesses RuvC domain which contributes to its single-strand break on target NA. Another difference between Cas12 and Cas9 is that Cas12 does not require Transactivating RNA and recognizes the target based on T-rich PAM sequence. An example of a biosensing platform that uses Cas12 is one-HOur Low-cost Multipurpose highly Efficient System (HOLMES). This system harnesses the cleavage activity of Cas12 with a quenched fluorescent to detect target DNA [84].

Integration of nanotechnology in biosensing technologies has been shown to improve sensitivity. The study conducted by Lee et al. (2021) developed a nanobiosensor for the detection of DENV. The system is designed based on Cas12 and methylene blue (MB) conjugated gold nanoparticles (MB-AuNPs) which increases the electrochemical factor. The performance evaluation of the electrochemical-based CRISPR biosensor exhibited 100 fM ultra-sensitive detection of DENV. One advantage of this system over several existing CRISPR-based biosensors is that it does not need the amplification step.

Wang et al. (2021) [85] developed a CRISPR-based nucleic acid detection platform known as Loop-mediated Isothermal Amplification coupled with CRISPR/Cas12a-mediated diagnosis (LACD) for the detection of *Mycobacterium tuberculosis*. The LACD assay comprised a LAMP amplification of the target DNA. The platform harnessed the trans-cleavage activity of Cas12a which cleaved target DNA. The degraded target DNA can be measured using a real-time fluorescence device or it can be visualized using a lateral flow biosensor. Evaluation of the sensitivity of the platform has shown that it can detect templates down to 50 fg of *Mycobacterium tuberculosis* Complex (MTC) genomic template per test.

Zhao et al. (2018) [86] developed an on-site biosensing technique for the detection of HIV. The sensor was designed based on hybridization between guide RNA coupled with Cas12a and target RNA which is amplified using Real-time Isothermal Reverse-transcription Recombinase-aided Amplification (rRT-RAA). The resulting cleavage of the target can be observed with the naked eye by using a blue light imager. Testing of the developed biosensing assay using clinical assay has shown that the system is capable of detecting 20 copies of purified HIV-1 RNA or DNA per reaction as low as 123 copies/mL of HIV-1 viral load.

The study conducted by Li et al., 2019 [87] developed Cas12b-based biosensor known as HOLMESV2. The biosensing platform harnessed the trans-cleavage collateral activity of the Cas system against target NA. The platform has shown excellent results in terms of discriminating single nucleotide polymorphism, detection of viral NA, human mRNA, and circular RNA. In order to avoid cross contamination and to amplify the target, the platform is designed along LAMP assay under constant temperature.

5.3. CRISPR/Cas13-Based Biosensors

Another remarkable discovery occurred in 2016 when Cas13 was discovered as a result of comprehensive research on the type VI CRISPR/Cas system. As discussed earlier, Cas9 possesses both RuvC and HNH domains and Cas12 possesses only RuvC domain. However, unlike both Cas12 and Cas9, Cas13 possesses special domains known as 2 Higher Eukaryotic and Prokaryotic Nucleotide (HEPN)-binding domains. Unlike Cas9, Cas13 possesses a special feature known as "collateral cleavage". This special feature of Cas13 can be harnessed to cut label RNA reporters for the detection of nucleic acid from a different target which includes viruses, bacteria, and eukaryotic cells [88].

Gootenberg et al. (2017) [89] developed a platform known as Specific High-Sensitivity Enzymatic Reporter UnLocking (SHERLOCK) which harnessed the collateral cleavage activity of Cas13a to detect specific strains of DENV and ZIKV. The platform is also capable of distinguishing pathogenic bacteria as well as identifying mutations in cell-free tumor DNA and genotype human DNA. Gootenberg et al. (2018) [90] developed a second version of SHERLOCK known as SHERLOCKv2, as a form of paper-based biosensing system for the detection of Zika virus RNA. The system is designed using guide RNA and Cas13a which recognizes the target and triggers collateral cleavage activity. The biosensor was able to achieve a detection limit as low as 20 aM. In order to address the need of the POC diagnostic platform for the detection of Ebola virus, Qin et al. (2019) [91] developed an automatic system which harnessed the collateral cleavage activity of Cas13a for degradation of target RNA. The degraded RNA fragments are measured using a custom fluorometer. The developed biosensing platform was able to achieve a result within 5 min and 20 pfu (5.45×10^7 copies/mL) detection limit. The use of CRISPR/Cas12 and Cas13 for detection of diseases is illustrated in Figure 4. Figure 5 shows the CRISPR system used to detect dengue fever using Cas12a/cpf1, with a target RNA DENV. The summary of Cas systems used for the detection of tropical diseases are shown in Table 3.

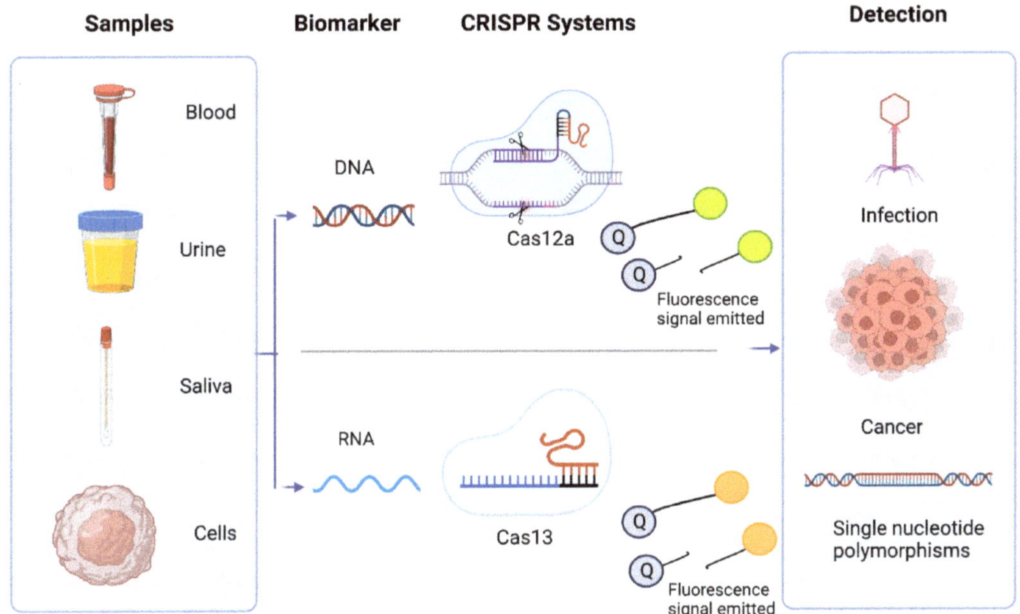

Figure 4. The use of CRISPR/Cas12 and Cas13 for detection of tropical diseases. (Adapted from Ref. [92] with free permission. Created with BioRender.com accessed on 21 September 2022).

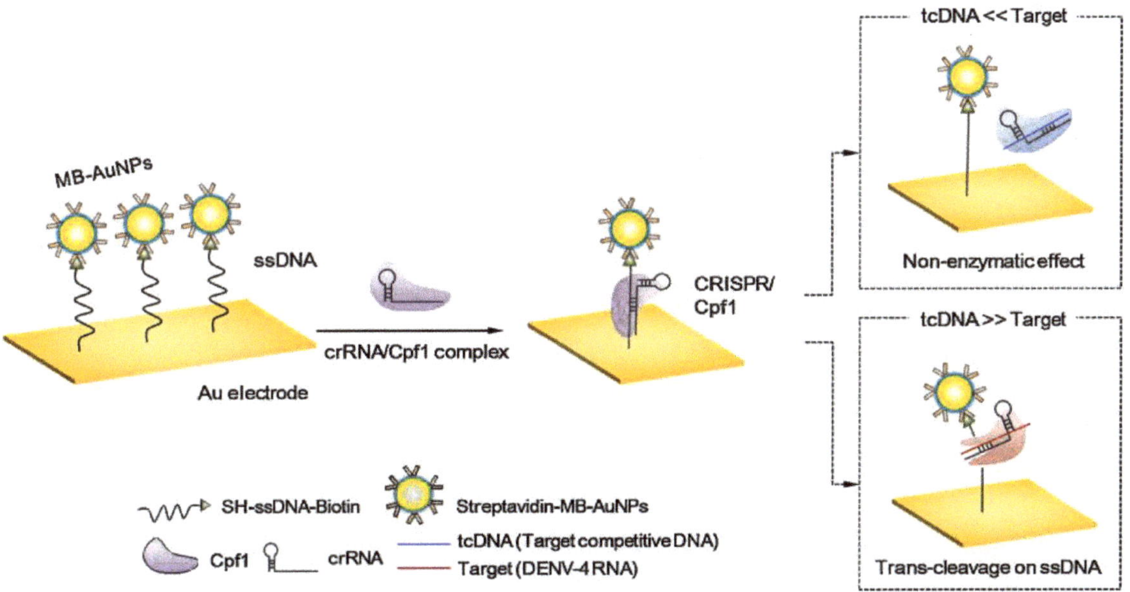

Figure 5. The CRISPR system for detection of RNA Dengue Virus (DENV). (Reprinted with permission from Elsevier [93]. Copyright and Licensing.).

Table 3. Detection of tropical diseases using the CRISPR/Cas system.

Cas System	Pathogen	References
Cas9/dCas9	*Mycobacterium tuberculosis*	[81]
	Zika and dengue virus	[82]
Cas12	Dengue virus	[94]
	Mycobacterium tuberculosis	[85]
	HIV	[86]
	Viral NA	[87]
Cas13	Dengue and Zika viruses	[89]
	Zika virus	[90]
	Ebola virus	[91]

6. Open Research Issue

The field of biosensing technology has witnessed progressive advancement in the past few years. What started merely as an electrochemical glucose biosensor has now been developed and transformed to molecular diagnosis of pathogenic disease. The main players or contributors to this transformation and innovations include the discovery of new approaches such as the CRISPR/Cas system, NA amplification techniques, nanotechnology, electronics, and material science. Despite the progress made so far in terms of the development of biosensors that function without amplifications steps, exhibition of high specificity based on SNP, and sensitivity of pM, fM, and aM concentrations, the biosensor field is still hindered by several challenges.

The current COVID-19 pandemic and the past Ebola, dengue, and Zika virus endemic have changed the landscape of clinical diagnosis from bench-lab assay to POC diagnostics. Scientists have proposed theoretical approaches and developed models and prototypes as well as a few POC devices for real-time diagnosis. Despite this progress, the developments of ideal, portable, cheap, precise, accurate, highly specific, and sensitive POC diagnostics biosensors still remain a challenge. The advancement in the field of computer science, the Internet of things (IoT), and Artificial Intelligence (AI) has opened the gateway to smart biosensors capable of collecting, storing, analyzing, and sharing data generated from biosensors in the form of numerical values or signals. However, smart biosensing technology is hampered by privacy and security issues which need to be addressed before it can be fully adopted into medical diagnosis.

7. Conclusions

TDs cause by DENV, ZIKV, Ebola, HIV, tuberculosis, etc., have caused havoc worldwide. Real-time, accurate, and early diagnoses of TDs is crucial for early treatment and epidemiological surveillance. Despite the wide array of clinical diagnostic approaches, including antibody, whole cell, and enzymatic-based techniques, nucleic acid-based detection approaches still remain the most sensitive and specific. The use of the RT-PCR-based approach has proved to be more efficient than antigen antibody-based methods due to its high specificity (hybridization) and amplification of target DNA.

The recent discovery of CRISPR/Cas systems in bacteria and archaea is revolutionizing the field of gene editing technology and biosensors. Several Cas systems have been identified and isolated from bacteria and programmed along with synthetic guide RNA to navigate through a long thread of genome in order to recognize a matching sequence. Scientists harness this activity in order to edit gene (insertion or deletion) for the development GMO and treatment of diseases. Several CRISPR-based biosensors have been developed including Cas9 and dCas9-based, Cas12-based, and Cas-13-based for the clinical diagnosis of bacterial and viral pathogens.

CRISPR/Cas system-based gene editing technology remains the most viable approach for eliminating inheritable diseases such as sickle cell anemia, Duchene muscular dystrophy (DMD), cystic fibrosis, Huntington's disease, etc. The biomimetic application of the system on editing target NA has open a Pandora's box for numerous futuristic applications on fighting diseases, enhancing features (such as designing babies who are immune to disease, increasing intelligence, enhancing eye color, etc.), and the detection of disease. Despite the hype of this technology, it is clouded by several challenges such as off-target the need for amplifications, the conversion of signals into readable output or numerical values, sensitivity beyond the femtomolar range, and ethical concerns. Thus, these challenges need to be addressed in order for this technology to reach its full potential.

Author Contributions: Conceptualization, Y.W.H.; validation, M.O. and S.G.; writing—original draft preparation, S.N.Z. and A.U.I.; writing—review and editing, S.N.Z., A.U.I. and M.O.; visualization, M.S.B. and M.I.H.L.Z.; supervision, Y.W.H. and M.O.; funding acquisition, Y.W.H., S.N.Z. and A.U.I. have the equal contribution. All authors have read and agreed to the published version of the manuscript.

Funding: This research was funded by Universitas Padjadjaran Academic Leadership Grant grant number 2203/UN6.3.1/PT.00/2022 and BUPP grant number 2203/UN6.3.1/PT.00/2022.

Institutional Review Board Statement: Not applicable.

Informed Consent Statement: Not applicable.

Data Availability Statement: Not applicable.

Acknowledgments: The support for this study was provided by Universitas Padjadjaran Academic Leadership Grant and BUPP Scheme No. 2203/UN6.3.1/PT.00/2022 and is gratefully acknowledged.

Conflicts of Interest: There are no conflict of interest to declare.

List of Abbreviations

Abbreviations	Meaning
A	Adenine
AchE	Acetylcholinesterase
AI	Artificial Intelligence
AIDS	Acquired Immune Deficiency Syndrome
aM	Attomolar
AuNPs	Gold Nanoparticles
Bp	Base Pair
C	Cytosine
CAS-EXARP	CRISPR/Cas9 Triggered Isothermal Exponential Amplification
CD4	Clusters of Differentiation 4
CDC	Center for Disease Control and Prevention
COVID-19	Coronavirus Disease 2019
CRISPR	Clustered Regularly Interspaced Short Palindromic Repeat
CrRNA	CRISPR RNA
CT	Computerized Tomography
CV	Cyclic Voltammetry
DNA	Deoxyribonucleic Acid
dCas9	Deactivated Cas9
DENV	Dengue Virus
DRC	Democratic Republic of Congo
DSB	Double-Strand Break
dsDNA	Double-Strand DNA
fM	Femtomolar
GMO	Genetically Modified Organism
HDR	Homologous Direct Repair
HEPN	Higher Eukaryotic and Prokaryotic Nucleotide

HOLMES	one-HOur Low-cost Multipurpose highly Efficient System
HIV	Human Immunodeficiency Virus
HR	Homologous Repair
HRP	Horseradish Peroxidase
IoT	Internet of Things
LAMP	Loop-Mediated Isothermal Amplification
LOD	Limit of Detection
KDA	Kilodalton
NA	Nucleic Acid
NASBA	Nucleic Acid Sequence-based Amplification
NASBACC	Nucleic Acid Sequence-based Amplification (NASBA)-CRISPR Cleavage
NHEJ	Non-homologous End-joining
NHPs	Non-Human Primates
NT	Nucleotide
NTDs	Non-Tropical Diseases
NTS	Non-Target Strand
MB	Methylene Blue
MTC	*Mycobacterium Tuberculosis* Complex
PAM	Protospacer Adjacent Motif
PC	Paired Cas9
PCR	Polymerase Chain Reaction
PFS	Protospacer Flanking Sequence
Pfu	Plague-forming Unit
Pm	Picomolar
POC	Point-of-Care
POCT	Point-of-Care Testing
RCA	Rolling Circle Amplification
rRT-RAA	Real-time Isothermal Reverse-transcription Recombinase-aided Amplification
RT-PCR	Real-Time Polymerase Chain Reaction
RNA	Ribonucleic Acid
SgRNA	Single Guide RNA
SHERLOCK	Specific High-Sensitivity Enzymatic Reporter UnLocking
SSDNA	Single-Stranded DNA
SSRNA	Single-Stranded RNA
TALENS	Transcription Activator-like Effector Nucleases
TacrRNA	Transactivating CRISPR RNA
TDs	Tropical Diseases
UNICEF	United Nation International Children's Emergency Fund
μM	Micromolar
U	Uracil
USA	United State of America
TB	Toluidine Blue
TMB	Tetramethylbenzidine
ZFN	Zinc-finger Nucleases
ZIKV	Zika Virus

References

1. Feasey, N.; Wansbrough-Jones, M.; Mabey, D.C.W.; Solomon, A.W. Neglected Tropical Diseases. *Br. Med. Bull.* **2010**, *93*, 179–200. [CrossRef] [PubMed]
2. Zumla, A.; Ustianowski, A. Tropical Diseases. Definition, Geographic Distribution, Transmission, and Classification. *Infect. Dis. Clin. N. Am.* **2012**, *26*, 195–205. [CrossRef]
3. Reegan, A.D.; Ceasar, S.A.; Paulraj, M.G.; Ignacimuthu, S.; Al-Dhabi, N.A. Current Status of Genome Editing in Vector Mosquitoes: A Review. *BioSci. Trends* **2016**, *10*, 424–432. [CrossRef] [PubMed]
4. Souza, A.A.; Ducker, C.; Argaw, D.; King, J.D.; Solomon, A.W.; Biamonte, M.A.; Coler, R.N.; Cruz, I.; Lejon, V.; Levecke, B.; et al. Diagnostics and the Neglected Tropical Diseases Roadmap: Setting the Agenda for 2030. *Trans. R. Soc. Trop. Med. Hyg.* **2021**, *115*, 129–135. [CrossRef]
5. Mahfouz, M.M.; Piatek, A.; Stewart, C.N. Genome Engineering via TALENs and CRISPR/Cas9 Systems: Challenges and Perspectives. *Plant Biotechnol. J.* **2014**, *12*, 1006–1014. [CrossRef]

6. Ibrahim, A.U.; Özsöz, M.; Saeed, Z.; Tirah, G.; Gideon, O. Genome Engineering Using the CRISPR Cas9 System. *Afr. J. Tradit. Complement. Altern. Med.* **2019**, *2*, 1000127. [CrossRef]

7. Ibrahim, A.U.; Al-Turjman, F.; Sa'id, Z.; Ozsoz, M. Futuristic CRISPR-Based Biosensing in the Cloud and Internet of Things Era: An Overview. *Multimed. Tools Appl.* **2020**, *81*, 35143–35171. [CrossRef] [PubMed]

8. McVeigh, P.; Maule, A.G. Can CRISPR Help in the Fight against Parasitic Worms? *eLife* **2019**, *8*, e44382. [CrossRef]

9. Tidman, R.; Abela-Ridder, B.; de Castañeda, R.R. The Impact of Climate Change on Neglected Tropical Diseases: A Systematic Review. *Trans. R. Soc. Trop. Med. Hyg.* **2021**, *115*, 147–168. [CrossRef]

10. Arnold, B.F.; van der Laan, M.J.; Hubbard, A.E.; Steel, C.; Kubofcik, J.; Hamlin, K.L.; Moss, D.M.; Nutman, T.B.; Priest, J.W.; Lammie, P.J. Measuring Changes in Transmission of Neglected Tropical Diseases, Malaria, and Enteric Pathogens from Quantitative Antibody Levels. *PLoS Negl. Trop. Dis.* **2017**, *11*, e0005616. [CrossRef]

11. LaBeaud, A.D. Why Arboviruses Can Be Neglected Tropical Diseases. *PLoS Negl. Trop. Dis.* **2008**, *2*, e247. [CrossRef]

12. Leong, A.S.Y.; Wong, K.T.; Leong, T.Y.M.; Tan, P.H.; Wannakrairot, P. The Pathology of Dengue Hemorrhagic Fever. *Semin. Diagn. Pathol.* **2007**, *24*, 227–236. [CrossRef] [PubMed]

13. Wu, W.; Bai, Z.; Zhou, H.; Tu, Z.; Fang, M.; Tang, B.; Liu, J.; Liu, L.; Liu, J.; Chen, W. Molecular Epidemiology of Dengue Viruses in Southern China from 1978 to 2006. *Virol. J.* **2011**, *8*, 322. [CrossRef] [PubMed]

14. Roy, S.K.; Bhattacharjee, S. Dengue Virus: Epidemiology, Biology, and Disease Aetiology. *Can. J. Microbiol.* **2021**, *67*, 687–702. [CrossRef] [PubMed]

15. Nyaruaba, R.; Mwaliko, C.; Mwau, M.; Mousa, S.; Wei, H. Arboviruses in the East African Community Partner States: A Review of Medically Important Mosquito-Borne Arboviruses. *Pathog. Glob. Health* **2019**, *113*, 209–228. [CrossRef] [PubMed]

16. Mwaliko, C.; Nyaruaba, R.; Zhao, L.; Atoni, E.; Karungu, S.; Mwau, M.; Lavillette, D.; Xia, H.; Yuan, Z. Zika Virus Pathogenesis and Current Therapeutic Advances. *Pathog. Glob. Health* **2021**, *115*, 21–39. [CrossRef] [PubMed]

17. Song, B.H.; Yun, S.I.; Woolley, M.; Lee, Y.M. Zika Virus: History, Epidemiology, Transmission, and Clinical Presentation. *J. Neuroimmunol.* **2017**, *308*, 50–64. [CrossRef]

18. Bhargavi, B.S.; Moa, A. Global Outbreaks of Zika Infection by Epidemic Observatory (EpiWATCH), 2016–2019. *Glob. Biosecurity* **2020**, *2*. [CrossRef]

19. Quaresma, J.A.S.; Pagliari, C.; Medeiros, D.B.A.; Duarte, M.I.S.; Vasconcelos, P.F.C. Immunity and Immune Response, Pathology and Pathologic Changes: Progress and Challenges in the Immunopathology of Yellow Fever. *Rev. Med. Virol.* **2013**, *23*, 305–318. [CrossRef] [PubMed]

20. Liu, W.; Sun, F.J.; Tong, Y.G.; Zhang, S.Q.; Cao, W.C. Rift Valley Fever Virus Imported into China from Angola. *Lancet Infect. Dis.* **2016**, *16*, 1226. [CrossRef]

21. Hamlet, A.; Gaythorpe, K.A.M.; Garske, T.; Ferguson, N.M. Seasonal and Inter-Annual Drivers of Yellow Fever Transmission in South America. *PLoS Negl. Trop. Dis.* **2021**, *15*, e0008974. [CrossRef]

22. Gaythorpe, K.A.M.; Hamlet, A.; Cibrelus, L.; Garske, T.; Ferguson, N.M. The Effect of Climate Change on Yellow Fever Disease Burden in Africa. *eLife* **2020**, *9*, e55619. [CrossRef]

23. Mukherjee, A.; Chawla-Sarkar, M. Rotavirus Infection: A Perspective on Epidemiology, Genomic Diversity and Vaccine Strategies. *Indian J. Virol.* **2011**, *22*, 11–23. [CrossRef]

24. Glass, R.I.; Tate, J.E.; Jiang, B.; Parashar, U. The Rotavirus Vaccine Story: From Discovery to the Eventual Control of Rotavirus Disease. *J. Infect. Dis.* **2021**, *224*, S331–S342. [CrossRef] [PubMed]

25. Luchs, A.; Timenetsky, M.D.C.S.T. Group A Rotavirus Gastroenteritis: Post-Vaccine Era, Genotypes and Zoonotic Transmission. *Einstein* **2016**, *14*, 278–287. [CrossRef] [PubMed]

26. Rosenberg, A.Z.; Naicker, S.; Winkler, C.A.; Kopp, J.B. HIV-Associated Nephropathies: Epidemiology, Pathology, Mechanisms and Treatment. *Nat. Rev. Nephrol.* **2015**, *11*, 150–160. [CrossRef] [PubMed]

27. Lucas, S.; Nelson, A.M. HIV and the Spectrum of Human Disease. *J. Pathol.* **2015**, *235*, 229–241. [CrossRef] [PubMed]

28. Kimmel, P.L.; Barisoni, L.; Kopp, J.B. Pathogenesis and Treatment of HIV-Associated Renal Diseases: Lessons from Clinical and Animal Studies, Molecular Pathologic Correlations, and Genetic Investigations. *Ann. Intern. Med.* **2003**, *139*, 214–226. [CrossRef]

29. Sullivan, P.S.; Johnson, A.S.; Pembleton, E.S.; Stephenson, R.; Justice, A.C.; Althoff, K.N.; Bradley, H.; Castel, A.D.; Oster, A.M.; Rosenberg, E.S.; et al. Epidemiology of HIV in the USA: Epidemic Burden, Inequities, Contexts, and Responses. *Lancet* **2021**, *397*, 1095–1106. [CrossRef]

30. Troncoso, A. Ebola Outbreak in West Africa: A Neglected Tropical Disease. *Asian Pac. J. Trop. Biomed.* **2015**, *5*, 255–259. [CrossRef]

31. Wadoum, R.E.G.; Sevalie, S.; Minutolo, A.; Clarke, A.; Russo, G.; Colizzi, V.; Mattei, M.; Montesano, C. The 2018–2020 Ebola Outbreak in the Democratic Republic of Congo: A Better Response Had Been Achieved Through Inter-State Coordination in Africa. *Risk Manag. Healthc. Policy* **2021**, *14*, 4923. [CrossRef] [PubMed]

32. Center for Disease Control and Prevention (CDC). CDC—Neglected Tropical Diseases—Diseases. Available online: https://www.cdc.gov/globalhealth/ntd/diseases/index.html (accessed on 22 August 2022).

33. Abebe, E.; Gugsa, G.; Ahmed, M. Review on Major Food-Borne Zoonotic Bacterial Pathogens. *J. Trop. Med.* **2020**, *2020*, 4674235. [CrossRef] [PubMed]

34. Churchyard, G.; Kim, P.; Shah, N.S.; Rustomjee, R.; Gandhi, N.; Mathema, B.; Dowdy, D.; Kasmar, A.; Cardenas, V. What We Know about Tuberculosis Transmission: An Overview. *J. Infect. Dis.* **2017**, *216*, S629–S635. [CrossRef] [PubMed]

35. Dheda, K.; Gumbo, T.; Maartens, G.; Dooley, K.E.; McNerney, R.; Murray, M.; Furin, J.; Nardell, E.A.; London, L.; Lessem, E.; et al. The Epidemiology, Pathogenesis, Transmission, Diagnosis, and Management of Multidrug-Resistant, Extensively Drug-Resistant, and Incurable Tuberculosis. *Lancet Respir. Med.* **2017**, *5*, 291–360. [CrossRef]

36. Stine, O.C.; Morris, J.G., Jr. Circulation and Transmission of Clones of *Vibrio cholerae* During Cholera Outbreaks. *Curr. Top. Microbiol. Immunol.* **2014**, *379*, 181–193. [CrossRef]

37. Baddam, R.; Sarker, N.; Ahmed, D.; Mazumder, R.; Abdullah, A.; Morshed, R.; Hussain, A.; Begum, S.; Shahrin, L.; Khan, A.I.; et al. Genome Dynamics of *Vibrio cholerae* Isolates Linked to Seasonal Outbreaks of Cholera in Dhaka, Bangladesh. *mBio* **2020**, *11*, 1–14. [CrossRef]

38. Reethy, P.S.; Lalitha, K.V. Characterization of *V. Cholerae* O1 Biotype El Tor Serotype Ogawa Possessing the CtxB Gene of the Classical Biotype Isolated from Well Water Associated with the Cholera Outbreak in Kerala, South India. *J. Water Health* **2021**, *19*, 478–487. [CrossRef]

39. Mwaba, J.; Debes, A.K.; Murt, K.N.; Shea, P.; Simuyandi, M.; Laban, N.; Kazimbaya, K.; Chisenga, C.; Li, S.; Almeida, M.; et al. Three Transmission Events of *Vibrio cholerae* O1 into Lusaka, Zambia. *BMC Infect. Dis.* **2021**, *21*, 570. [CrossRef] [PubMed]

40. Ameer, M.A.; Wasey, A.; Salen, P. *Escherichia coli (E Coli 0157 H7)*; StatPearls Publishing: Treasure Island, FL, USA; Tampa, FL, USA, 2021. Available online: https://europepmc.org/article/NBK/nbk507845 (accessed on 22 August 2022).

41. Bonten, M.; Johnson, J.R.; van den Biggelaar, A.H.J.; Georgalis, L.; Geurtsen, J.; De Palacios, P.I.; Gravenstein, S.; Verstraeten, T.; Hermans, P.; Poolman, J.T. Epidemiology of *Escherichia coli* Bacteremia: A Systematic Literature Review. *Clin. Infect. Dis.* **2021**, *72*, 1211–1219. [CrossRef]

42. Hotez, P.J. *Human Parasitology and Parasitic Diseases: Heading Towards 2050*, 1st ed.; Elsevier Ltd.: Amsterdam, The Netherlands, 2018; Volume 100. ISBN 9780128151693.

43. MacKintosh, C.L.; Beeson, J.G.; Marsh, K. Clinical Features and Pathogenesis of Severe Malaria. *Trends Parasitol.* **2004**, *20*, 597–603. [CrossRef]

44. Autino, B.; Corbett, Y.; Castelli, F.; Taramelli, D. Pathogenesis of Malaria in Tissues and Blood. *Mediterr. J. Hematol. Infect. Dis.* **2012**, *4*, e2012061. [CrossRef] [PubMed]

45. Escalante, A.A.; Pacheco, M.A. Malaria Molecular Epidemiology: An Evolutionary Genetics Perspective. *Microbiol. Spectr.* **2019**, *7*, 1–18. [CrossRef]

46. Al-Awadhi, M.; Ahmad, S.; Iqbal, J. Current Status and the Epidemiology of Malaria in the Middle East Region and Beyond. *Microorganisms* **2021**, *9*, 338. [CrossRef] [PubMed]

47. Villar, J.C.; Perez, J.G.; Cortes, O.L.; Riarte, A.; Pepper, M.; Marin-Neto, J.A.; Guyatt, G.H. Trypanocidal Drugs for Chronic Asymptomatic *Trypanosoma cruzi* Infection. *Cochrane Database Syst. Rev.* **2014**, 1–63. [CrossRef] [PubMed]

48. Kioy, D.; Jannin, J.; Mattock, N. Human African Trypanosomiasis. *Nat. Rev. Microbiol.* **2004**, *2*, 186–187. [CrossRef] [PubMed]

49. MacLean, L.M.; Odiit, M.; Chisi, J.E.; Kennedy, P.G.E.; Sternberg, J.M. Focus-Specific Clinical Profiles in Human African Trypanosomiasis Caused by *Trypanosoma brucei rhodesiense*. *PLoS Negl. Trop. Dis.* **2010**, *4*, 1–12. [CrossRef]

50. World Health Organization. Trypanosomiasis, Human African (Sleeping Sickness). Available online: https://www.who.int/news-room/fact-sheets/detail/trypanosomiasis-human-african-(sleeping-sickness) (accessed on 22 August 2022).

51. Steverding, D. The History of Leishmaniasis. *Parasites Vectors* **2017**, *10*, 82. [CrossRef]

52. Scott, P.; Novais, F.O. Cutaneous Leishmaniasis: Immune Responses in Protection and Pathogenesis. *Nat. Rev. Immunol.* **2016**, *16*, 581–592. [CrossRef] [PubMed]

53. Tripathi, L.K.; Nailwal, T.K. Leishmaniasis: An Overview of Evolution, Classification, Distribution, and Historical Aspects of Parasite and Its Vector. In *Pathogenesis, Treatment and Prevention of Leishmaniasis*; Academic Press: Cambridge, MA, USA, 2021; pp. 1–25. [CrossRef]

54. Hatherell, H.A.; Simpson, H.; Baggaley, R.F.; Hollingsworth, T.D.; Pullan, R.L. Sustainable Surveillance of Neglected Tropical Diseases for the Post-Elimination Era. *Clin. Infect. Dis.* **2021**, *72*, S210–S216. [CrossRef] [PubMed]

55. Kappagoda, S.; Ioannidis, J.P.A. La Prevención y El Control de Enfermedades Tropicales Desatendidas: Una Visión General de Ensayos Aleatorios, Exámenes Sistemáticos y Metaanálisis. *Bull. World Health Organ.* **2014**, *92*, 356–366. [CrossRef]

56. Waddington, S.N.; Privolizzi, R.; Karda, R.; O'Neill, H.C. A Broad Overview and Review of CRISPR-Cas Technology and Stem Cells. *Curr. Stem Cell Rep.* **2016**, *2*, 9–20. [CrossRef]

57. Chylinski, K.; Makarova, K.S.; Charpentier, E.; Koonin, E.V. Classification and Evolution of Type II CRISPR-Cas Systems. *Nucleic Acids Res.* **2014**, *42*, 6091–6105. [CrossRef]

58. Barrangou, R.; Fremaux, C.; Deveau, H.; Richardss, M.; Boyaval, P.; Moineau, S.; Romero, D.A.; Horvath, P. CRISPR Provides Against Viruses in Prokaryotes. *Science* **2007**, *315*, 1709–1712. [CrossRef] [PubMed]

59. Garneau, J.E.; Dupuis, M.È.; Villion, M.; Romero, D.A.; Barrangou, R.; Boyaval, P.; Fremaux, C.; Horvath, P.; Magadán, A.H.; Moineau, S. The CRISPR/Cas Bacterial Immune System Cleaves Bacteriophage and Plasmid DNA. *Nature* **2010**, *468*, 67–71. [CrossRef]

60. Dronina, J.; Bubniene, U.S.; Ramanavicius, A. The Application of DNA Polymerases and Cas9 as Representative of DNA-Modifying Enzymes Group in DNA Sensor Design (Review). *Biosens. Bioelectron.* **2021**, *175*, 112867. [CrossRef] [PubMed]

61. Gustafsson, C. *A Tool for Genome Editing*; The Royal Swedish Academy of Sciences: Stockholm, Sweden, 2020; pp. 1–14.

62. Zhang, Y.; Wu, Y.; Wu, Y.; Chang, Y.; Liu, M. CRISPR-Cas Systems: From Gene Scissors to Programmable Biosensors. *Trends Anal. Chem.* **2021**, *137*, 116210. [CrossRef]

63. Swarts, D.C.; Jinek, M. Cas9 versus Cas12a/Cpf1: Structure—Function Comparisons and Implications for Genome Editing. *Wiley Interdiscip. Rev. RNA* **2018**, *9*, e1481. [CrossRef]

64. O'Connell, M.R. Molecular Mechanisms of RNA Targeting by Cas13-Containing Type VI CRISPR—Cas Systems. *J. Mol. Biol.* **2019**, *431*, 66–87. [CrossRef]

65. Harrington, L.B.; Burstein, D.; Chen, J.S.; Paez-Espino, D.; Ma, E.; Witte, I.P.; Cofsky, J.C.; Kyrpides, N.C.; Banfield, J.F.; Doudna, J.A. Programmed DNA Destruction by Miniature CRISPR-Cas14 Enzymes. *Science* **2018**, *362*, 839–842. [CrossRef]

66. Doudna, J.A.; Charpentier, E. The New Frontier of Genome Engineering with CRISPR-Cas9. *Science* **2014**, *346*, 1–11. [CrossRef]

67. Du, X.; McManus, D.P.; French, J.D.; Jones, M.K.; You, H. CRISPR/Cas9: A New Tool for the Study and Control of Helminth Parasites. *BioEssays* **2021**, *43*, 2000185. [CrossRef]

68. Grzybek, M.; Golonko, A.; Górska, A.; Szczepaniak, K.; Strachecka, A.; Lass, A.; Lisowski, P. The CRISPR/Cas9 System Sheds New Lights on the Biology of Protozoan Parasites. *Appl. Microbiol. Biotechnol.* **2018**, *102*, 4629–4640. [CrossRef]

69. Dong, Y.; Simões, M.L.; Marois, E.; Dimopoulos, G. CRISPR/Cas9-Mediated Gene Knockout of *Anopheles gambiae* FREP1 Suppresses Malaria Parasite Infection. *PLoS Pathog.* **2018**, *14*, e1006898. [CrossRef] [PubMed]

70. Macias, V.M.; McKeand, S.; Chaverra-Rodriguez, D.; Hughes, G.L.; Fazekas, A.; Pujhari, S.; Jasinskiene, N.; James, A.A.; Rasgon, J.L. Cas9-Mediated Gene-Editing in the Malaria Mosquito Anopheles Stephensi by ReMOT Control. *G3 Genes Genomes Genet.* **2020**, *10*, 1353–1360. [CrossRef]

71. Zhang, C.; Li, Z.; Cui, H.; Jiang, Y.; Yang, Z.; Wang, X. Systematic CRISPR-Cas9-Mediated Modifications of *Plasmodium yoelii* ApiAP2 Genes Reveal Functional Insights into Parasite Development. *mBio* **2017**, *8*, 1986–2003. [CrossRef] [PubMed]

72. Rostami, M.N. CRISPR/Cas9 Gene Drive Technology to Control Transmission of Vector-Borne Parasitic Infections. *Parasite Immunol.* **2020**, *42*, e12762. [CrossRef]

73. Doerflinger, M.; Forsyth, W.; Ebert, G.; Pellegrini, M.; Herold, M.J. CRISPR/Cas9—The Ultimate Weapon to Battle Infectious Diseases? *Cell. Microbiol.* **2017**, *19*, e12693. [CrossRef]

74. Hammond, A.; Galizi, R.; Kyrou, K.; Simoni, A.; Siniscalchi, C.; Katsanos, D.; Gribble, M.; Baker, D.; Marois, E.; Russell, S.; et al. A CRISPR-Cas9 Gene Drive System Targeting Female Reproduction in the Malaria Mosquito Vector *Anopheles gambiae*. *Nat. Biotechnol.* **2016**, *34*, 78–83. [CrossRef] [PubMed]

75. Burle-Caldas, G.A.; Soares-Simões, M.; Lemos-Pechnicki, L.; DaRocha, W.D.; Teixeira, S.M.R. Assessment of Two CRISPR-Cas9 Genome Editing Protocols for Rapid Generation of *Trypanosoma Cruzi* Gene Knockout Mutants. *Int. J. Parasitol.* **2018**, *48*, 591–596. [CrossRef] [PubMed]

76. Zhang, W.W.; Matlashewski, G. CRISPR-Cas9-Mediated Genome Editing in Leishmania Donovani. *mBio* **2015**, *6*, 1–14. [CrossRef] [PubMed]

77. Sollelis, L.; Ghorbal, M.; Macpherson, C.R.; Martins, R.M.; Kuk, N.; Crobu, L.; Bastien, P.; Scherf, A.; Lopez-Rubio, J.J.; Sterkers, Y. First Efficient CRISPR-Cas9-Mediated Genome Editing in Leishmania Parasites. *Cell. Microbiol.* **2015**, *17*, 1405–1412. [CrossRef] [PubMed]

78. Beneke, T.; Madden, R.; Makin, L.; Valli, J.; Sunter, J.; Gluenz, E. A CRISPR Cas9 High-Throughput Genome Editing Toolkit for Kinetoplastids. *R. Soc. Open Sci.* **2017**, *4*, 1–16. [CrossRef] [PubMed]

79. Li, Y.; Li, S.; Wang, J.; Liu, G. CRISPR/Cas Systems towards Next-Generation Biosensing. *Trends Biotechnol.* **2019**, *37*, 730–743. [CrossRef] [PubMed]

80. Charpentier, E.; Doudna, J.A. A Nobel Prize for Genetic Scissors. *Nat. Mater.* **2021**, *20*, 1. [CrossRef]

81. Zhang, Y.; Wang, Y.; Xu, L.; Lou, C.; Ouyang, Q.; Qian, L. Paired DCas9 Design as a Nucleic Acid Detection Platform for Pathogenic Strains. *Methods* **2022**, *203*, 70–77. [CrossRef]

82. Pardee, K.; Green, A.A.; Takahashi, M.K.; Braff, D.; Lambert, G.; Lee, J.W.; Ferrante, T.; Ma, D.; Donghia, N.; Fan, M.; et al. Rapid, Low-Cost Detection of Zika Virus Using Programmable Biomolecular Components. *Cell* **2016**, *165*, 1255–1266. [CrossRef]

83. Qiu, X.Y.; Zhu, L.Y.; Zhu, C.S.; Ma, J.X.; Hou, T.; Wu, X.M.; Xie, S.S.; Min, L.; Tan, D.A.; Zhang, D.Y.; et al. Highly Effective and Low-Cost MicroRNA Detection with CRISPR-Cas9. *ACS Synth. Biol.* **2018**, *7*, 807–813. [CrossRef]

84. Li, S.Y.; Cheng, Q.X.; Li, X.Y.; Zhang, Z.L.; Gao, S.; Cao, R.B.; Zhao, G.P.; Wang, J.; Wang, J.M. CRISPR-Cas12a-Assisted Nucleic Acid Detection. *Cell Discov.* **2018**, *4*, 18–21. [CrossRef]

85. Wang, Y.; Li, J.; Li, S.; Zhu, X.; Wang, X.; Huang, J.; Yang, X.; Tai, J. LAMP-CRISPR-Cas12-Based Diagnostic Platform for Detection of *Mycobacterium tuberculosis* Complex Using Real-Time Fluorescence or Lateral Flow Test. *Microchim. Acta* **2021**, *188*, 347. [CrossRef]

86. Zhao, S.; Wang, S.; Zhang, S.; Liu, J.; Dong, Y. State of the Art: Lateral Flow Assay (LFA) Biosensor for on-Site Rapid Detection. *Chin. Chem. Lett.* **2018**, *29*, 1567–1577. [CrossRef]

87. Li, L.; Li, S.; Wu, N.; Wu, J.; Wang, G.; Zhao, G.; Wang, J. HOLMESv2: A CRISPR-Cas12b-Assisted Platform for Nucleic Acid Detection and DNA Methylation Quantitation. *ACS Synth. Biol.* **2019**, *8*, 2228–2237. [CrossRef]

88. East-Seletsky, A.; O'Connell, M.R.; Knight, S.C.; Burstein, D.; Cate, J.H.D.; Tjian, R.; Doudna, J.A. Two Distinct RNase Activities of CRISPR-C2c2 Enable Guide-RNA Processing and RNA Detection. *Nature* **2016**, *538*, 270–273. [CrossRef]

89. Gootenberg, J.S.; Abudayyeh, O.O.; Lee, J.W.; Essletzbichler, P.; Dy, A.J.; Joung, J.; Verdine, V.; Donghia, N.; Daringer, N.M.; Freije, C.A.; et al. Nucleic Acid Detection with CRISPR-Cas13a/C2c2. *Science* **2017**, *356*, 438–442. [CrossRef]

90. Gootenberg, J.S.; Abudayyeh, O.O.; Kellner, M.J.; Joung, J.; Collins, J.J.; Zhang, F. Multiplexed and Portable Nucleic Acid Detection Platform with Cas13, Cas12a and Csm6. *Science* **2018**, *360*, 439–444. [CrossRef] [PubMed]

91. Qin, P.; Park, M.; Alfson, K.J.; Tamhankar, M.; Carrion, R.; Patterson, J.L.; Griffiths, A.; He, Q.; Yildiz, A.; Mathies, R.; et al. Rapid and Fully Microfluidic Ebola Virus Detection with CRISPR-Cas13a. *ACS Sens.* **2019**, *4*, 1048–1054. [CrossRef] [PubMed]
92. Lou, J.; Wang, B.; Li, J.; Ni, P.; Jin, Y.; Chen, S.; Xi, Y.; Zhang, R.; Duan, G. The CRISPR-Cas System as A Tool for Diagnosing and Treating Infectious Diseases. *Mol. Biol. Rep.* **2022**, 1–11. [CrossRef]
93. Lee, Y.; Choi, J.; Han, H.K.; Park, S.; Park, S.Y.; Park, C.; Baek, C.; Lee, T.; Min, J. Fabrication of Ultrasensitive Electrochemical Biosensor for Dengue Fever Viral RNA Based on CRISPR/Cpf1 Reaction. *Sens. Actuators B Chem.* **2021**, *326*, 128677. [CrossRef]
94. Lee, R.A.; De Puig, H.; Nguyen, P.Q.; Angenent-Mari, N.M.; Donghia, N.M.; McGee, J.P.; Dvorin, J.D.; Klapperich, C.M.; Pollock, N.R.; Collins, J.J. Ultrasensitive CRISPR-Based Diagnostic for Field-Applicable Detection of *Plasmodium* Species in Symptomatic and Asymptomatic Malaria. *Proc. Natl. Acad. Sci. USA* **2020**, *117*, 25722–25731. [CrossRef] [PubMed]

Tropical Medicine and Infectious Disease

Article

The Impact of Flooding on Snail Spread: The Case of Endemic Schistosomiasis Areas in Jiangxi Province, China

Shang-Biao Lv [1,†], Ting-Ting He [1,†], Fei Hu [1], Yi-Feng Li [1], Min Yuan [1], Jing-Zi Xie [1], Zong-Guang Li [1], Shi-Zhu Li [2] and Dan-Dan Lin [1,3,*]

[1] Jiangxi Provincial Institute of Parasitic Diseases, Nanchang 330096, China
[2] National Institute of Parasitic Diseases, China CDC (Chinese Center for Tropical Diseases Research), Key Laboratory on Parasite and Vector Biology, National Health Commission, WHO Collaborating Centre for Tropical Diseases, National Center for International Research on Tropical Diseases, Ministry of Science and Technology, Shanghai 200025, China
[3] Jiangxi Province Key Laboratory of Schistosomiasis Prevention and Control, Nanchang 330096, China
* Correspondence: jxlindandan@163.com
† These authors contributed equally to this work.

Abstract: Flooding is the main natural factor in snail diffusion, and it has a negative impact on schistosomiasis transmission. There are few studies on the spread and migration of snails following a flood; therefore, we aimed to investigate the influence of flooding on snail diffusion and explore the characteristics and laws of snail diffusion in Jiangxi Province. By using a retrospective survey and cross-sectional survey, the data on snail spreading in Jiangxi Province from 2017 to 2021 were collected. The distribution, nature, and area of snail spread were systematically analyzed in combination with the hydrological situation, types of region, and types of flood. From 2017 to 2021, a total of 120 snail-spread environments were found, including in 92 hilly areas and in 28 lake areas. The areas caused by flood and by other means numbered 6 and 114, respectively. The proportions of recurrence, expansion, and first-time occurrences were 43.42%, 38.16%, and 18.42%, respectively, and the 14 new snail environments were only distributed in the hilly areas. With the exception of 2018, the ratio of snail-spread areas in the hilly region was higher than that in lake region in other years. The average density of live snails was 0.0184–1.6617 no./0.1 m^2 and 0.0028–0.2182 no./0.1 m^2 in the hilly region. Among the 114 environments affected by floods, 86 consisted of hilly environments, including 66 spreading environments affected by rainstorm floods, and 20 rainstorm debris flow environments. There were 28 lake areas, of which 10 were in the Jiangxi section of Yangtze River and were affected by rainstorm floods. Snail spread following flooding has a certain 'lag effect,' and = simple annual changes in hydrological characteristics have little effect on the diffusion of snails or on their density = in the affected environment, but it is more closely related to local floods. The hilly environments are more susceptible to floods than the lake region, and the risk of snail spread is much higher in the hilly than in the lake region.

Keywords: flood; snail spread; schistosomiasis; Jiangxi province

Citation: Lv, S.-B.; He, T.-T.; Hu, F.; Li, Y.-F.; Yuan, M.; Xie, J.-Z.; Li, Z.-G.; Li, S.-Z.; Lin, D.-D. The Impact of Flooding on Snail Spread: The Case of Endemic Schistosomiasis Areas in Jiangxi Province, China. *Trop. Med. Infect. Dis.* **2023**, *8*, 259. https://doi.org/10.3390/tropicalmed8050259

Academic Editor: Vyacheslav Yurchenko

Received: 13 March 2023
Revised: 28 April 2023
Accepted: 28 April 2023
Published: 30 April 2023

1. Introduction

Oncomelania hupensis is the only intermediate host of Schistosoma japonicum in China. Where schistosomiasis is endemic, Oncomelania hupensis is found in abundance. Snail control is one of the effective measures to block the transmission of or eliminate schistosomiasis [1–3]. In China, according to the geographical environment, snail habitats are classified into three types, lakes, hills, and water networks, and these are mainly distributed in the marshland in the middle and lower reaches of Yangtze River, the marshland of Dongting Lake and Poyang Lake, and in hilly ditches, fields, ponds, and other environments in Yunnan, Sichuan, Guangxi, and Zhejiang [4]. Over the past 70 years, various measures, such as molluscicides, environmental modifications, and integrated management, have been used to control snail populations, and the rate of schistosomiasis epidemics has dropped to its lowest levels in history in snail-infested areas. The size of snail-infested areas has decreased from 14.8 billion square meters when schistosomiasis control first began to 3.69 billion square meters in 2021, a decrease of 97.56%. Among the regions affected, the lake, hilly, and water-network regions were 3497.37 km^2, 191.81 km^2, and 3.50 km^2 in size, respectively [5,6]. Although the snail areas have been effectively controlled, new and reemerging environments develop every year in endemic schistosomiasis provinces across the country. From 2017 to 2021, the total size of newly infested snail areas nationwide reached 2571.77 km^2, of which five lake-region provinces accounted for 98.79% [7–9]. In addition, from 2009 to 2017, a total of 59.9 million m^2 of new snail infestations were found downstream of the Three Gorges dam, accounting for 93.70% of the total newly infested snail area [10–13].

Oncomelania hupensis is a mollusk, with a distribution, reproduction, and growth that are influenced by natural factors, such as floods, climate change, environmental changes, and human and animal activities. Social factors, including water-conservation projects, flood control, levee construction, wetland protection, and farming methods, also influence their distribution, reproduction, and growth [4]. Water is one of the necessary conditions for snail growth and reproduction. Locations with fluctuating water levels or slow water flow and vegetation growth are often suitable habitats for snails. Studies have shown that changes in water levels greatly affect the distribution of oncomelanid snails, and floods are extreme expressions of water level changes [14].

Jiangxi Province used to be one of the provinces in China with a relatively severe prevalence of schistosomiasis. In 2015, the province reached a standard level of control over the transmission of schistosomiasis. As of the end of 2021, twenty-four out of thirty-nine counties in the province where schistosomiasis is prevalent had reached an elimination standard, six counties had reached a transmission-interrupted standard, and nine counties had maintained transmission-control standards [11]. However, because the snail-breeding environment has not been completely changed, especially the water situation of winter land and summer water in the Poyang Lake area, infestations are difficult to control artificially; snails still breed, reproduce, and spread in the Poyang Lake area, along the beaches of Yangtze River, and in some hilly ditches, wasteland, and paddy fields in northeast and central Jiangxi. The snail areas in Jiangxi Province have always ranked second in the whole country, reaching 849.38 km^2, 23.00% of China's total infested area, with the snail habitats divided into hilly and lake types. In recent years, the snail area in Jiangxi Province has been on the rise [6,15,16]. It is obvious that the breeding, reproduction, and spread of snails are increasingly becoming a huge risk in schistosomiasis control.

In China, numerous studies have shown that floods have a negative impact on schistosomiasis transmission. However, there are few studies that use the characteristics of floods as complex systems to analyze the spread and migration of snails [17–20]. Therefore, in this paper, the distribution, nature and areas of snail diffusion in Jiangxi Province from 2017 to 2021 were systematically analyzed in combination with the hydrological situation, type of region, and flood invasion methods, in order to explore the characteristics of snail spread in Jiangxi Province and provide a scientific basis for the formulation of snail-control measures in Jiangxi Province and even the whole country.

2. Method

2.1. Snail Spreading

2.1.1. Data Source

Based on the snail-census results in 2016 [16], the snail diffusion in historical snail environments or in snail-free environments adjacent to snail-occupied environments in endemic schistosomiasis areas over the whole province was investigated. A combination of retrospective and current situation surveys was used to review and collect the investigation data of snails in endemic schistosomiasis counties from 2017 to 2020. Additionally, to analyze the impact of the catastrophic flood disaster in Jiangxi province in 2020 on snail spread in endemic schistosomiasis areas, a special snail-diffusion survey was carried out in the affected endemic areas in the spring of 2021, and the area surveyed was suspected to be a snail-breeding environment and a historical snail environment affected by floods. The snail survey was conducted according to the industry-standard survey of oncomelanid snails (WS/T563-2017) using a snail frame measuring 0.1 m^2, according to the standard specified by the Ministry of Health in China.

2.1.2. Spreading-Environment Types

In accordance with snails' natural habits, the spreading environment used in the study was divided into a hilly region and a lake region, and the snail indicators, such as the snail area, average density of live snails, and their occurrence rate (%) in snail frames were counted for each spreading environment.

2.1.3. Spreading Properties

Based on the results of the snail-proliferation survey and historical information, the diffusion properties were divided into expansion environment (expansion), recurrence environment (recurrence), and new environment (new). The expansion environment is an environment in which the snail area is expanded due to natural factors in the snail-free environment, to which the original snail distribution is connected. The recurrence environment is an environment in which snails are confirmed to be extinct after successful control in a historical snail environment, and live snails have been found again several years later. The newly discovered snail environment refers to the environment where snails are detected for the first time in an environment in which they had not been observed before.

2.2. Flooding

2.2.1. Water Information

Checking the water-resources bulletin of Jiangxi Province, water information, such as hydrological data of the Poyang Lake basin, annual rainfall of the province, and the extent of flooding in endemic schistosomiasis areas since 2016 were collected. According to the annual precipitation or annual runoff, the annual hydrological characteristics from 2016 to 2021 are classified as wet year, median-water year, or dry year [21–23].

2.2.2. Flood-Invasion Manner

Based on the local flooding data, the way in which each snail-spread environment was flooded was divided into the following manner [24,25].

Storm-flood type: A flood formed by heavy rain fall with a greater-than-normal intensity, through the production and confluence of flows in ditches, rivers, and other streams. The main characteristics of the flood are the high peak volume, long duration, and wide flooding area.

Storm-debris-flow type: A flood that contains a large amount of sediment, clay, gravel, rock, and other types of solid material in hilly areas, which are mixed with stormwater to make the gully area move or flow and slide slowly down the slope of the gully.

Submerged-lake type: A flood causing the water level of Poyang Lake to rise rapidly due to the joint influence of the top backflow of Yangtze River and the flooding of Ganjiang, Fei, Xinjiang, Rao, and Xiu Rivers into Poyang Lake, causing the beach of Poyang Lake to become inundated by the flood for a long period of time.

2.3. Data Analysis

Two snail indicators, the amount of space occupied by live snails in the snail frame and calculated as percentage of frames with live snails and the mean density of live snails, were studied using a descriptive statistical analysis. The qualitative data were compared using a chi-squared test (χ^2) test and the quantitative data were analyzed using a variance analysis. A linear regression analysis was used to compare the snail-diffusion-area ratio and annual change trend. All data were statistically analyzed with SPSS 20.0 software (IBM, Armonk, NY, USA). Any $p < 0.05$ indicated that the difference was statistically significant.

$$\text{Percentage of frames with live snail } (\%) = \frac{\text{frames of live snails}}{\text{Number of investigation frames}} \times 100\%$$

$$\text{Mean density of live snail } (\text{No.} / 0.1 \, \text{m}^2) = \frac{\text{number of live snails}}{\text{Number of investigation frames}} \tag{1}$$

3. Results

3.1. Analysis of the Oncomelania Snail Spread

From 2017 to 2021, a total of 120 snail-spread environments were found, with varying degrees of occurrence in all years except 2018. Among these, the numbers of hilly and marshland regions were ninety-two and twenty-eight, respectively, of which six were not caused by floods. There was a significant difference in the number of snail-spread environments between the flooded and non-flooded regions ($\chi^2 = 194.40$, $p = 0.00$). In addition, the number of spreading environments in the hilly areas was greater than that in the lake region for all years except 2017 (Figure 1).

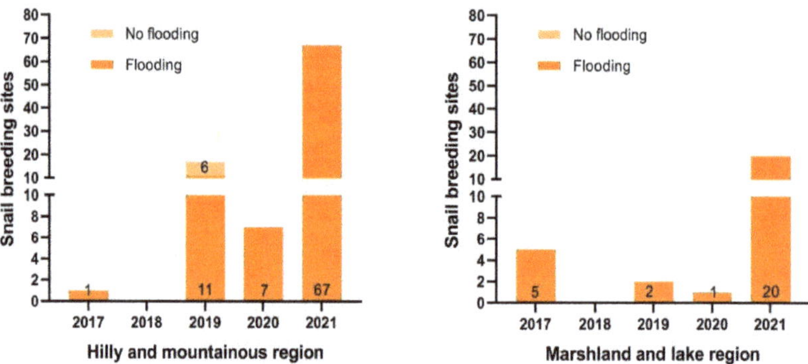

Figure 1. The annual distribution of the number of snail-spread environments in different endemic areas.

3.1.1. Distribution of the Snail-Spread Environment

The 114 snail-spread environments caused by floods were mainly distributed in the hilly areas of northeastern Jiangxi, followed by Yangtze River and Poyang Lake. In particular, six environments were distributed in the hilly area of northeastern Jiangxi and Yangtze River in 2017, thirteen and eight environments were distributed in the hilly areas of northeastern Jiangxi and the river channel of Poyang Lake in 2019 and 2020, respectively, and eighty-seven environments were distributed in the hilly areas of northeastern Jiangxi, Yangtze River, and Poyang Lake in 2021. Furthermore, the beach elevation of Poyang Lake was 16.1–17.9 m (Wusong benchmark), and the beach elevation in the estuary of the Poyang Lake inlet was 10.0–12.5 m.

3.1.2. Snail-Spread Area

The spread-area results from 2017–2021 showed that the hilly areas' spread ratio was greater than that of the lake region in each year except for 2018, and that it increased from year to year ($R^2 = 0.854$, F = 17.615, $p = 0.025$). In the lake region, the area ratio remained generally stable from 2017 to 2020, and increased rapidly by 1.49% in 2021, but there was no significant trend from year to year ($R^2 = 0.51$, F = 3.12, $p = 0.18$) (Table 1).

Table 1. Spreading areas of Oncomelania hupensis in different endemic regions from 2017 to 2021.

Year	Hilly and Mountainous Region			Marshland and Lake Region		
	Actual Snail-Infested Area Total (km²)	Area with Snails Detected for the First Time (km²)	Proportion (%)	Actual Snail Infested-Area Total (km²)	Area with Snails Detected for the First Time (km²)	Proportion (%)
2017	24.31	0.02	0.09	809.28	0.59	0.07
2018	24.30	0	0	809.83	0	0
2019	25.22	0.31	1.24	810.08	0.52	0.06
2020	25.29	0.78	3.08	810.63	0.55	0.07
2021	27.41	1.81	6.59	821.97	12.21	1.49

3.1.3. Snails' Spreading Properties

In 2017–2021, the proportions of snail recurrence, expansion, and first-time occurrences were 43.42%, 38.16%, and 18.42%, respectively. A pairwise comparison showed that the recurrence rate was significantly higher than the first-time-occurrence rate ($\chi^2 = 11.119$, $p = 0.001$), and was not statistically different from the expansion rate ($\chi^2 = 0.436$, $p = 0.509$). The expansion was significantly higher than the first-time occurrence ($\chi^2 = 7.297$, $p = 0.007$). Snail recurrence and expansion were present in both hilly and lake environments, and the 14 new environments were only found in the hilly region. The mean density of live snails was highest in the expansion environment, followed by the new environment and, next, by the recurring environment, while the percentage equating to the amount of space occupied by live snails in the snail frames was highest in the new environment, followed by the expansion environment and the recurrence (Table 2).

Table 2. The composition of Oncomelania hupensis' spreading properties.

Property	Number of Diffusion Environments		Area (km²)	The Mean Density of Live Snails (No./0.1 m²)	The Space Occupied by Live Snails (%)
	Hilly and Mountainous Region	Marshland and Lake Region			
Recurrence	39	23	14.19	0.20	11.74
Expansion	33	5	2.35	0.62	33.90
First-time occurrence	14	0	0.12	0.53	37.19
Total	86	28	16.66	0.46	22.92

3.1.4. Snail Situation

The mean density of live snails in the spreading environments in the hilly and lake regions was 0.018–1.66 no./0.1 m² and 0.0028–0.22 no./0.1 m², respectively, with a median value of 0.58 no./0.1 m for the hilly region, which was significantly higher than the median value of 0.039 no./0.1 m for the lake region ($p = 0.00$).

The space occupied by live snails in the frames was 1.21–69.44% and 0.28–6.40% in the hilly and lake regions, respectively. The median percentage of the frames with snails in the hilly region was 36.15%, which was significantly higher than that of the lake region (2.07%) ($p = 0.000$) (Figure 2).

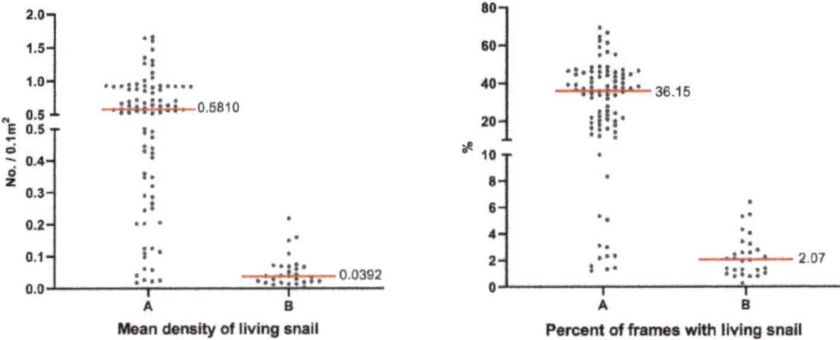

Figure 2. The Oncomelania hupensis' situation in the diffusion environments of different endemic regions.

A refers to the hilly and mountainous regions and B refers to the marshland and lake regions.

3.2. Impact of the Flooding Patterns on the Spread of Snails

The annual-rainfall-analysis results showed that 2016 and 2019 were wet years, 2017 and 2020 were median water years, and 2018 was a dry year. Regardless of the type of year, disasters with different degrees of flooding occur in some areas of Jiangxi Province every year. From 2016 to 2019, the flood disasters in the endemic area were mainly concentrated in the hilly areas of northeastern Jiangxi and the Jiangxi section of Yangtze River, and only short-term flooding above the warning level occurred in Poyang Lake. In July 2020, the Poyang Lake area was doubly affected by the flood's top-down irrigation from the five rivers and Yangtze River, which was hit by the catastrophic flood in 1998. It rose at a rate of more than 0.4 m for 8 consecutive days, with a maximum daily increase of 0.65 m, varying by nearly 7 m from its lowest to its highest. The water levels at the Xingzi, Poyang, Yongxiu, and Guxiandu hydrological stations exceeded the historical water levels by 0.11 m, 0.14 m, 0.15 m, and 0.25 m, respectively. All 185 single retreat dikes in the Poyang Lake area were used for flood diversion and inflow for the first time since they were built, in 2007, on July 11, and the historical or suspected snail environment in the flood-detention basin was submerged (Table 3).

Table 3. Precipitation and the hydrological characteristics in Jiangxi Province from 2016 to 2020.

Year	Average Precipitation in the Province (mm)	Comparison with the Multi-Year Average (mm)	Type
2016	1997.0	21.9	wet year
2017	1637.2	5.9	median water year
2018	1129.9	−26.9	dry year
2019	2032.7	31.5	wet year
2020	1666.7	7.8	median water year

Among the 114 environments affected by flooding, there were 86 diffuse environments in the hilly areas, including 66 spreading environments affected by storm flooding and 20 by storm-debris flow. There were twenty-eight lake and marsh areas, of which ten areas in the Jiangxi section of Yangtze River were affected by storm flooding, fifteen were affected in the main lake area of Poyang Lake, and three were affected in the river channel, which was affected by the lake inundation.

According to the flood-invasion types, the average density of live snails and the percentage of occupied space in the frames with live snails, from high to low, were the rainstorm-flood type (0.50 No./0.1 m^2 and 22.84 %), rainstorm-debris-flow type (0.33 No./0.1 m^2 and 14.31%), and submerged-lake type (0.06 No./0.1 m^2 and 2.16%). There were significant differences in the percentages of occupied space in the frames with live snails between the two groups (χ^2 = 206.97, 2668.38, 1001.83, P$_{all}$ = 0.00). However, the differences in the mean density of live snails were only statistically significant between the rainstorm-flooding and submerged-lake types (F = 20.60, *p* = 0.00). There was no significant difference between the rainstorm-flood type and the rainstorm-debris-flow type, nor between the rainstorm-debris-flow type and the submerged-lake type (F = 1.38, 27.73; *p* = 0.339, 0.06) (Table 4).

Table 4. Effect of different types of flood invasion on snail spread.

Type	Number of Diffusion Environments		Area (km^2)	The Mean Density of Live Snails (No./0.1 m^2)	The Space Occupied by Live Snails (%)
	Hilly and Mountainous Regions	Marshland and Lake Regions			
Storm flood	66	10	4.70	0.58	36.48
Storm-debris flow	20	0	2.55	0.23	12.77
Submerged lake	0	18	11.71	0.04	2.15
Total	86	28	16.66	0.46	22.92

4. Discussion

The "Healthy China 2030" planning outline proposes that the whole country will achieve the goal of eliminating schistosomiasis by 2030 [26]. *Oncomelania hupensis* is the only intermediate host of Schistosoma japonicum in China, and controlling the snail population plays a vital role in blocking and eliminating schistosomiasis. Jiangxi Province is one of the provinces with a severe schistosomiasis prevalence and is also one of the main provinces with frequent flooding in China. At present, the snail areas in Jiangxi Province cover 849.38 km^2, and they are mainly distributed in Poyang Lake, along Yangtze River, and in the beach environments to which the water system is connected, as well as in low-lying, slow-flowing, and overgrown irrigation ditches and paddy fields, dry land, ponds, weirs, hillside wasteland, and other environments among the hills in the northeast and central–southwest areas of Jiangxi [16]. Numerous studies have shown that flood disasters are one of the important factors causing the resurgence of snail populations [26–28]. The results of this study also verified this point. The number of diffusion environments caused by floods was 12.67 times that of non-floods.

The results showed that 75.44% of the snail-spread environments were located in the hilly region after flooding; furthermore, the area ratio of the snail spread, the percentage of occupied space in the frames with live snails, and the average density of live snails in the hilly region were significantly higher than those in the lake region. It is suggested that hilly areas are more susceptible to flooding than lake areas; therefore, the risk of snails spreading in hilly areas is greater than in other areas and, due to the complexity of this environment and its proximity to villages, residents and livestock are in contact with water more frequently due to production or living; thus, they are more at risk. However, the spread area of snails in the lake areas reached 13.88 km^2 after flooding, which was nearly five times that of the hilly areas, mainly because of the flat terrain and single vegetation; the areas with snails in the lake region increased sharply after flooding. In the hilly region, the distribution of Oncomelania snails was relatively isolated, often appearing as small units that were not connected to each other along the water system, with small areas of snails, spotty patterns of distribution, and complex and diverse environments [29].

New snail areas often appear in hilly regions, which are closely related to the complex environments and crisscross water systems. Floods are likely to cause oncomelanid snail colonization, breeding, and reproduction in adjacent environments, to which water systems are connected (or interlinked). In lake regions, flooding can easily lead to the reappearance of or increase in snails due to flat beach areas. In addition, the area where snails were found for the first time was in the adjacent non-endemic zones that were not frequently included in the annual snail-survey plan and so were easily overlooked by investigators. They could only be detected when the density of snails was high, with a high occurrence rate in the snail frames. Snail-reproduction or -expansion environments often appear in historical snail environments, which are included in the local snail-monitoring range all year round. Therefore, these environments can often be discovered in time, so the density of live snails and the amount of space occupied in the frames by live snails is low.

This study showed that the occurrence area and the magnitude of flooding determined the spatial distribution and scope of snail spread [30–34]. From 2016 to 2019, in Jiangxi Province, the flood disasters were mainly concentrated in the hilly areas of northeastern Jiangxi and in the western section of Yangtze River. The flooded areas reflected the spatial distribution of the snail spread, which mainly encompassed the hills of northeastern Jiangxi, Yangtze River, and the watercourse of Poyang Lake into Yangtze River. No snail spread occurred in the main lake area of Poyang Lake.

In 2020, a catastrophic flood exceeding that which took place in 1998 occurred in the Poyang Lake basin, covering almost all of the snail-infected areas in Jiangxi Province. The snail-survey data in 2021 showed that snail spread occurred not only in the hilly areas of northeastern Jiangxi and Yangtze River, but also in the upper elevation area of the beach of the main lake area of Poyang Lake. This may have been related to the migration mode of oncomelanid snails, which can be dispersed through suspended mass, transported mass or by water surface-transport, which is affected by the environmental water depth, flow velocity, and floating objects [34].

For example, local floods can easily cause snails in hilly areas and on the Yangtze River beach to spread in the mass-transfer mode downstream along the surrounding water system. Moreover, the constant water level of Poyang Lake makes it difficult for snails to spread to the upper elevation of the beach with the movement of suspended sediment, but when floods recede, the snails can be moved with the suspended sediment to a lower elevation of the beach. Long-term super-warming water levels can cause oncomelanid snails to move upward with suspended sediments to the higher elevation of marshlands in the lake area [35].

At the same time, the snail spread has no obvious relationship with simple annual hydrological characteristics, but it is closely related to flooding in regional areas, especially water-level changes. This study showed that the diffusion distribution of snails from 2016 to 2020 was consistent with the distribution of local flood disasters. Although 2020 was a median water year, due to the basin-wide large flood in Poyang Lake in July, which broke the previous water-level record, the number of diffusion environments in 2021 was significantly higher than that in other years, accounting for 72.50% of the total in the five years covered by this study and 76.31% of the total diffusion area. This also suggests that there was a "lag effect" of the water regime on the diffusion of Oncomelania snails in that year, and recurrence, expansion, or the appearance of newly snail-infected environments occurred only one year after the flood disaster, which was consistent with previous reports [36–39].

The survey data showed that different flood-invasion types have different effects on snail habitats. Rainstorm flooding and storm-debris flow mainly affect snail spread in hills, and the mean density of live snails and the percentage of occupied space in frames with live snails were higher than those in the submerged lake. The main reasons for these observations were as follows. First, a sudden catastrophic flood could have caused the snails to migrate long distances downstream with the water flow in mass-transfer mode, leading to the snails' redistribution in hills, ditches, paddy fields, and river beaches. Second, the storm-debris flow could easily have caused waterlogging and encouraged the migration and diffusion of snails or young snails to suitable surrounding environments in the form of suspended matter, and even led to the emergence of new snail spots in non-endemic villages.

For example, the newly discovered snail distribution, found in 2021, in a previously snail-free village (Sangyuan Village) in Yushan County was due to the catastrophic flooding disaster in 2020. Third, the groundwater and karst caves are rich in the endemic area of Jiangxi. The groundwater in these karst caves is swollen due to the large basin-wide floods, resulting in residual snails drifting with the groundwater flow, in the form of surface transportation, to the ground to recolonize and reproduce. For example, snails were captured 51 years after they were eradicated from an area in Licun village, Wuyuan County.

Lake submersion is the main factor affecting snail spread in the Poyang Lake area. Studies have shown that the water level affects the elevation distribution of snails in this type of environment. The duration of high water levels affects the range and distance of snail diffusion, and the velocity and direction of the water flow affect the snails' spreading modes and destinations [31,35].

In the flood season, due to the double impact of the incoming water from the five rivers and the top support of Yangtze River, the water level of Poyang Lake's backflow rises rapidly. Young or adult snails, along with floating objects, are affected by the water back-flow and reach the upper elevation of the beach, which has no snails. The higher the water level, the greater the distance of the lake's top-supported back-flow, resulting in higher elevation and, consequently, the distribution of the highest snail line in the lake area. For example, a total of 15 snail-recurrence environments in the main lake area of Poyang Lake were discovered in 2021, all of which were distributed on the north shore of Poyang Lake and at the tail of Ganjiang River, where the elevation is above 16.1 m.

In addition, the young snails or adult snails drift downstream, transported along the water surface during the rapid decline of the lake water after flooding, and the consequent lower water level causes the snails to spread to the lower elevations of the beaches. For example, in 2019 and 2020, there were three diffusion environments at the mouth of Poyang Lake, with an elevation below 12.5 m, indicating that some water-regime factors, such as the annual water-level changes, the periodic inundation, and the dew duration of the marshland, can also determine the nature of the snail diffusion, density, and distribution area on the submerged beaches of the lake.

Schistosomiasis control is a complex systematic project, and flood disasters make this prevention and control work more complex and arduous. Numerous studies have shown that the first-time occurrence and reoccurrence areas of snails after flooding disasters can reach 2.16 and 2.56 times those in normal years, respectively; they may affect the snail distribution over the subsequent 3–5 years, and may have a continuous impact on the prevalence of schistosomiasis [40–42]. Following 70 years of prevention and control, no newly infected individuals or livestock have been detected in Jiangxi Province since 2020. The number of existing schistosomiasis patients is no higher than 6000, and the schistosomiasis epidemic is at its lowest level in history. However, susceptible individuals, such as travelers and anglers, are still at risk, and awareness of prevention and control is quite weak, posing a potential risk of schistosomiasis epidemics in areas to which snails have spread. Therefore, in addition to strengthening health education and disease surveillance for susceptible populations, it is truly necessary to strengthen snail monitoring in disaster-affected areas to eliminate snail distribution and spread in a timely fashion, to formulate precise mollusk-control programs according to the local conditions, and deal with newly discovered snail spots as soon as possible, so as to prevent the further spread of snails caused by floods.

In hilly regions, in addition to strengthening the measures of snail control, coordination and cooperation between departments should be strengthened. Projects, such as the structural adjustment of the planting and breeding industries, the transformation of high-standard fertile land, the restoration of barren fields and replanting, and environmental improvements should be given priority to cover snail areas and, consequently, completely change the snail-infected environment. In lake regions, the focus should be on strengthening the implementation of infection-source-control measures, such as prohibiting grazing in snail-infected marshland, replacing cattle with machines, carrying out drug spraying to kill snails or larvae in high-risk areas with schistosomes, and promptly repairing water-damaged projects and facilities to prevent the post-disaster rebound of endemic schistosomiasis [25,28].

5. Conclusions

Snail spread is one of the main reasons for schistosomiasis transmission, and flooding disasters are the main factors causing snail spread. This study has shown that the dispersal of oncomelanid snails after floods has a certain "lag effect," and that simple annual changes in hydrological characteristics have no obvious effect on the dispersal of Oncomelania or on the snail density in the dispersal environment, which is more closely related to local floods. Hilly areas are more susceptible to floods than the lakes, the number of diffusion environments and the snail-diffusion-area ratio in these areas are greater than in lake areas, and the risk of snail proliferation is also much higher than in lake areas. Therefore, in the future, it is necessary to strengthen the monitoring of snail spread after flooding, to take control measures according to the local conditions to prevent snail spread, and to further strengthen the implementation of infection-source-control measures, such as the prohibition of grazing in snail-infected areas and the use of machines instead of bovines to prevent the rebounding of the schistosomiasis endemic. This research is of important guiding significance for guiding China's post-flood snail monitoring and to formulate snail-diffusion-control measures in the future.

Author Contributions: D.-D.L. conceived of and designed the research. F.H., S.-B.L., T.-T.H., Y.-F.L., Z.-G.L., M.Y. and J.-Z.X. performed the survey. S.-B.L., F.H. and T.-T.H. completed the analyses and co-wrote the paper. S.-Z.L. and D.-D.L. contributed to the writing and revisions. All authors have read and agreed to the published version of the manuscript.

Funding: This research was supported by the National Natural Science Foundation of China (grant number 81660557 and 71764011) and Jiangxi Province Focus on Research and Development Plan (grant number 20171BBG70105) and Key Laboratory Plan of Jiangxi Province (grant number 20192BCD40006). Natural Science Foundation of Jiangxi province (grant number 20212BAB206074).

Institutional Review Board Statement: Not applicable.

Informed Consent Statement: Not applicable.

Data Availability Statement: The data used in the study are available from the corresponding author upon reasonable request.

Acknowledgments: We thank the Jiangxi Provincial Schistosomiasis and Endemic Diseases Control Office for its administrative support and the technical staff at the Chinese Center for Diseases Control and Prevention and all 39 schistosomiasis-control institutions in Jiangxi province for their assistance in the fieldwork.

Conflicts of Interest: The authors declare that they have no competing interests. All authors have read and agreed to the published version of the manuscript.

References

1. Feng, J.X.; Gong, Y.F.; Luo, Z.W.; Wang, W.; Cao, C.L.; Xu, J.; Li, S.Z. Scientific basis of strategies for schistosomiasis control and prospect of the 14th Five-Year Plan in China. *Chin. J. Parasitol. Parasit. Dis.* **2022**, *40*, 428–435.
2. Yuan, Y.; Cao, C.L.; Huang, X.B.; Zhao, Q.P. Oncomelania hupensis control strategy during the stage moving towards elimination of schistosomiasis in China. *Chin. J. Schisto. Control* **2022**, *34*, 337–340+351.
3. Zheng, J.X. *Prediction on Transmission Risk of Schistosomiasis and Liver Flukes Diseases in China and Mekong River Basin*; Chinese Center for Disease Control and Prevention: Beijing, China, 2021.
4. Zhou, X.N. *Science on Oncomelania Snail*; Science Press: Beijing, China, 2005.
5. Xu, J.; Hu, W.; Yang, K.; Lv, S.; Li, S.Z.; Zhou, X.N. Key points and research priorities of schistosomiasis control in China during the 14th Five-Year Plan Period. *Chin. J. Schisto. Control* **2021**, *33*, 1–6.
6. Zhang, L.J.; Xu, Z.M.; Yang, F.; Dang, H.; Li, Y.L.; Lv, S.; Cao, C.L.; Xu, J.; Li, S.Z.; Zhou, X.N. Endemic status of schistosomiasis in People's Republic of China in 2020. *Chin. J. Schisto. Control* **2021**, *33*, 225–233.
7. Zhang, L.J.; Xu, Z.M.; Dai, S.W.; Dang, H.; Xu, J.; Li, S.Z.; Zhou, X.N. Endemic status of schistosomiasis in People's Republic of China in 2017. *Chin. J. Schisto. Control* **2018**, *30*, 481–488.
8. Zhang, L.J.; Xu, Z.M.; Guo, J.Y.; Dai, S.W.; Dang, H.; Lv, S.; Xu, J.; Li, S.Z.; Zhou, X.N. Endemic status of schistosomiasis in People's Republic of China in 2018. *Chin. J. Schisto. Control* **2019**, *31*, 576–582.
9. Zhang, L.J.; Xu, Z.M.; Dang, H.; Li, Y.L.; Lv, S.; Xu, J.; Li, S.Z.; Zhou, X.N. Endemic status of schistosomiasis in People's Republic of China in 2019. *Chin. J. Schisto. Control* **2020**, *32*, 551–558.
10. Cao, C.L.; Zhang, L.J.; Bao, Z.P.; Dai, S.M.; Xu, J.; Li, S.Z.; Zhou, X.N. Endemic situation of schistosomiasis in People's Republic of China from 2010 to 2017. *Chin. J. Schisto. Control* **2019**, *31*, 519–521+554.
11. Wang, Q.; Xu, J.; Zhang, L.J.; Zheng, H.; Ruan, Y.; Hao, Y.W.; Li, S.Z.; Zhou, X.N. Analysis of endemic changes of schistosomiasis in China from 2002 to 2010. *Chin. J. Schisto. Control* **2015**, *27*, 229–234+250.
12. Dai, S.M.; Edwards, J.K.; Guan, Z.; Lv, S.; Li, S.Z.; Zhang, L.J.; Feng, J.; Feng, N.; Zhou, X.N.; Xu, J. Change patterns of oncomelanid snail burden in areas within the Yangtze River drainage after the three gorges dam operated. *Infect. Dis. Poverty* **2019**, *8*, 48–56. [CrossRef]
13. Zhang, L.J.; Xu, Z.M.; Yang, F.; He, J.Y.; Dang, H.; Li, Y.L.; Cao, C.L.; Xu, J.; Li, S.Z.; Zhou, X.N. Progress of schistosomiasis control in People's Republic of China in 2021. *Chin. J. Schisto. Control* **2022**, *34*, 329–336.
14. Hu, F.; Ge, J.; Lv, S.B.; Li, Y.F.; Li, Z.J.; Yuan, M.; Chen, Z.; Liu, Y.M.; Li, Y.S.; Ross, A.G.; et al. Distribution pattern of the snail intermediate host of schistosomiasis japonica in the Poyang Lake region of China. *Infect. Dis. Poverty* **2019**, *29*, 8–23. [CrossRef] [PubMed]
15. Liu, Y.W.; Lin, D.D. *Investigation and Analysis Report on Snails in Schistosomiasis Endemic Areas in Jiangxi Province (2015–2017)*; Jiangxi Sci Technol Press: Nanchang, China, 2018.
16. Lv, S.B.; Li, Y.F.; Chen, Z.; Gu, X.N.; Yuan, M.; Hu, F.; Li, Z.J.; Lin, D.D. Study on distribution status of Oncomelania hupensis, the intermediate host of Schistosoma japonicum in Jiangxi Province II Spatial-temporal distribution of snail-infested environments. *Chin. J. Schisto. Control* **2018**, *30*, 396–403.
17. Zhang, S.Q.; Wang, T.P.; Ge, J.H.; Tao, C.G.; Zhang, G.H.; Lv, D.B.; Wu, W.Z.; Wang, Q.Z. Influence on the diffusion of snail by flooding in Anhui province. *J. Trop. Dis. Parasitol.* **2004**, *2*, 90–94.

18. Huang, Y.X.; Hong, Q.B.; Gao, Y.; Gao, Y.; Zhang, L.H.; Chen, H.; Guo, J.H.; Liang, Y.S. Impact of floodwater on schistosomiasis transmission in area of lower reaches of Yangtze River. *Chin. J. Schisto. Control* **2006**, *18*, 401–405.
19. Du, S.Q.; Wen, J.H. Challenges and suggestions for flood risk management in China from the flood season disaster situation in 2020. Chin. *J. Disast. Reduct.* **2020**, *380*, 12–15.
20. Cheng, H.G.; Lin, D.D.; Zhang, S.J.; Hu, F.; Ning, A.; Zeng, X.J.; Hu, G.H.; Xie, Z.W. studies on the influence of flood disaster on schistosomiasis transmission and on its control in Poyang lake region I. analysis of schistosomiasis prevalence in flood year and following year. *Chin. J. Schisto. Control.* **2001**, *13*, 141–146.
21. GB/T 22482-2008; Standard for Hydrological Information and Hydrological Forecasting. General Administration of Quality Supervision, Inspection and Quarantine of the People's Republic of China; Standardization Administration: Beijing, China, 2009.
22. *Investigation and Evaluation of Water Resources in Jiangxi Province*; Jiangxi Provincial Hydrology Bureau: Nanchang, China, 2004.
23. Zhu, H.T. *Set Pair Analysis Methods for Risk Analysis of Flood Disaster*; Heifei University of Technology: Heifei, China, 2010.
24. *Water Resources Bulletin of Jiangxi Province (2016–2021)*; Jiangxi Provincial Water Resources Department: Nanchang, China, 2021.
25. Li, B. *Tutorial for Outline of the Healthy China 2030 Plan*; Springer: Singapore, 2016.
26. Zhang, S.Q. Flood disasters and schistosomiasis control. *Chin. J. Schisto. Control* **2020**, *32*, 522–525.
27. Li, K.J.; Pi, Y.; Xiao, Y.; Yuan, Y.; Wang, H.; Liu, J.B. Effect evaluation of the preventive and control measures against schistosomiasis after the floods in Hubei province in 2020. *Chin. J. Disaster Med.* **2021**, *9*, 1216–1218+1247.
28. Cao, C.L.; Li, S.Z.; Zhou, X.N. Impact of schistosomiasis transmission by catastrophic flood damage and emergency response in China. *Chin. J. Schisto. Control* **2016**, *28*, 618–623.
29. Xue, J.Z.; Li, Y.F.; Yuan, M.; He, T.T.; Li, Z.G.; Lv, S.B.; Lin, D.D. Investigation on the diffusion of snails after flood disaster in 2020 in Jiangxi Province. *J. Trop. Dis. Parasitol.* **2022**, *20*, 197–201.
30. Jiang, T.T.; Yang, K. Progresses of research on patterns and monitoring approaches of Oncomelania hupensis spread. *Chin. J. Schisto. Control* **2020**, *32*, 208–211.
31. Lv, S.B.; Lin, D.D. Natural environment and schistosomiasis transmission in Poyang Lake region. *J. Schistosomiasis Control* **2014**, *26*, 561–564+574.
32. Ma, W.; Liao, W.G.; Feng, S.X. *Hydrological Impact Mechanism and Risk Assessment Method of Schistosomiasis Transmission*; China Water & Power Press: Beijing, China, 2011; pp. 57–58.
33. Lv, S.B.; Yuan, M.; Hu, F.; Lv, S.; Li, Y.F.; Hang, C.Q.; Lin, D.D. Distribution characteristic and the main influencing factors of Oncomelania hupensis in different years in Poyang Lake area of Jiangxi Province. *Chin. J. Schistosomiasis Control* **2019**, *37*, 637–643.
34. Zhang, L.; Wang, J.S.; Yang, Q.H.; Min, F.Y. Research of Oncomelania Snails'Migration and Diffusion in Water: Review and Prospects. *J. Yangtze River Sci. Res. Inst.* **2019**, *36*, 7–12.
35. Ma, W.; Liao, W.G.; Kuang, S.F.; Xiao, S.B.; Li, P.A. On correlation between diffusion of oncomelania hupensis gredler and the flow regime in Dongting lake. *Resour. Environ. Yangtze Basin* **2009**, *18*, 264–269.
36. Wang, J.S.; Lu, J.Y.; Wei, G.Y.; Yao, S.M. Impact of Environment Changes on Oncomelania Spread. *J. Yangtze River Sci. Res. Inst.* **2007**, *24*, 16–19.
37. Ma, W.; Liao, W.G.; Kuang, S.F.; Xiao, S.B. Study on the quantitative relationship of oncomelania hupensis diffusion with the flow regime of the Dongting Lake. *J. China Inst. Water Resour. Hydropower Res.* **2009**, *7*, 15–20.
38. Wang, H.; Shan, X.C.; Zhong, C.H.; Liu, S.; Yang, J.J.; Liu, J.B. Investigation on the diffusion of snails after flood disaster in 2020 in Hubei Province. *J. Trop. Dis. Parasitol.* **2022**, *20*, 202–205.
39. Xu, J. *Study on the Regulation of Water Conservancy and Schistosomiasis Control Measures in Dongting Lake Area Based on the Spatiotemporal Evolution of Comprehensive Risk of Schistosomiasis*; Hunan Normal University: Changsha, China, 2021.
40. Wu, X.; Zhang, S.Q.; Xu, X.J.; Hang, Y.X.; Steinmann, P.; Utzinger, J.; Wang, T.-P.; Xu, J.; Zheng, J.; Zhou, X.-N. Effect of floods on the transmission of schistosomiasis in the Yangtze River valley, People's Republic of China. *Parasitol. Int.* **2008**, *57*, 271–276. [CrossRef]
41. Zhang, L.J.; Zhu, H.Q.; Zhu, H.Q.; Wang, Q.; Lv, S.; Xu, J.; Li, S.Z. Assessment of schistosomiasis transmission risk along the Yangtze River basin after the flood disaster in 2020. *Chin. J. Schisto. Control* **2020**, *32*, 464–468+475.
42. Chen, Y.; Zhang, S.Q.; Gao, L.; Zhang, R. Influence of flood disaster on Oncomelania snail distribution in Wuhu section of Yangtze River in 2020. *China Trop. Med.* **2022**, *22*, 941–946.

Tropical Medicine and Infectious Disease

Article

Pathological Changes in Hepatic Sinusoidal Endothelial Cells in *Schistosoma japonicum*-Infected Mice

Tingting Jiang [1], Xiaoying Wu [1], Hao Zhou [1], Yuan Hu [1,*] and Jianping Cao [1,2,*]

[1] National Institute of Parasitic Diseases, Chinese Center for Disease Control and Prevention, (Chinese Center for Tropical Diseases Research), Key Laboratory of Parasite and Vector Biology, National Health Commission of People's Republic of China, World Health Organization Collaborating Center for Tropical Diseases, Shanghai 200025, China

[2] The School of Global Health, Chinese Center for Tropical Diseases Research, Shanghai Jiao Tong University School of Medicine, Shanghai 200025, China

* Correspondence: huyuan@nipd.chinacdc.cn (Y.H.); caojp@chinacdc.cn (J.C.)

Abstract: Schistosomiasis japonica is a zoonotic parasitic disease causing liver fibrosis. Liver sinusoidal endothelial cells (LSECs) exhibit fenestrations, which promote hepatocyte regeneration and reverses the process of liver fibrosis. To investigate the pathological changes of LSECs in schistosomiasis, we established a Schistosomiasis model. The population, phenotype, and secretory function of LSECs were detected by flow cytometry at 20, 28, and 42 days post infection. The changes in LSEC fenestration and basement membrane were observed through scanning electron microscopy (SEM) and transmission electron microscopy (TEM). Quantitative real-time PCR and Western blotting were used to detect the expression of molecules associated with epithelial–mesenchymal transition (EMT) and fibrosis of LSECs and the liver. The flow cytometry results showed that the total LSEC proportions, differentiated LSEC proportions, and nitric oxide (NO) secretion of LSECs were decreased, and the proportion of dedifferentiated LSECs increased significantly post infection. The electron microscopy results showed that the number of fenestrate was decreased and there was complete basement membrane formation in LSECs following infection. The qPCR and Western blot results showed that EMT, and fibrosis-related indicators of LSECs and the liver changed significantly during the early stages of infection and were aggravated in the middle and late stages. The pathological changes in LSECs may promote EMT and liver fibrosis induced by *Schistosoma japonicum* infection.

Keywords: *Schistosoma japonicum*; liver sinusoidal endothelial cells; de-differentiation; epithelial–mesenchymal transition; liver fibrosis

Citation: Jiang, T.; Wu, X.; Zhou, H.; Hu, Y.; Cao, J. Pathological Changes in Hepatic Sinusoidal Endothelial Cells in *Schistosoma japonicum*-Infected Mice. *Trop. Med. Infect. Dis.* **2023**, *8*, 124. https://doi.org/10.3390/tropicalmed8020124

Academic Editor: Vyacheslav Yurchenko

Received: 12 January 2023
Revised: 8 February 2023
Accepted: 9 February 2023
Published: 17 February 2023

1. Introduction

Schistosomiasis japonica is a zoonotic parasitic disease. It is primarily prevalent in China, the Philippines, and other Asian countries. Eggs are deposited in the host liver, causing portal hypertension, liver fibrosis, and splenomegaly, seriously threatening human health [1]. By the end of 2020, there were still 450 endemic counties (cities and districts) in China, 29,517 cases of advanced schistosomiasis, and approximately 71,370,400 people at risk of infection [2]. After several efforts, schistosomiasis infection has been well controlled in China. Infection control was achieved in 2008, and transmission control was completed in 2015 [3]. In contrast, the risk of transmission remains high due to several hosts and the ecological environment for snail breeding [4]. Schistosomiasis remains endemic in some areas of China [5]. Moreover, there are still many chronic schistosomiasis patients in China [2]. In February 2022, the WHO published new guidelines to update the global public health strategy against schistosomiasis [6]. Due to a lack of effective treatment for schistosomiasis liver fibrosis, there is an urgent need to study the regulatory mechanism of liver fibrosis.

Chronic liver injury leads to liver inflammation and fibrosis, which activates myofibroblasts and leads to the secretion of extracellular matrix proteins [7]. Liver sinusoidal endothelial cells (LSECs) in the hepatic sinuses are the first cells to respond to liver injury [8,9]. As a barrier between hepatocytes and blood flow, LSECs have fenestrations due to a lack of a basal membrane, which can promote nutrient transport [8]. Under physiological conditions, LSECs can regulate hepatic vascular tension and help maintain low portal pressure [9]. LSECs maintain hepatic stellate cells and Kupffer cells in the resting state and promote immune tolerance in the liver [8]. Previous research has shown that LSECs are involved in initiating hepatocyte regeneration and reversing the process of liver fibrosis. In contrast, abnormal activation of LSECs associated with chronic liver injury can also induce liver fibrosis [10].

There are two phenotypes of LSECs, including differentiated and dedifferentiated phenotypes. Differentiated LSECs have abundant fenestral structures and lack a complete basement membrane, which is essential to inducing hepatocyte regeneration and maintaining HSCs in a resting state [9]. Dedifferentiated LSECs lose their fenestral structure and form a complete basement membrane beneath the cells. This can activate HSC transformation into myofibroblasts and promote the development of liver fibrosis [11,12]. Some studies have shown that LSECs can induce the dedifferentiation type [13] or endothelial-to-mesenchymal transition (EMT) during the early stage of hepatic fibrosis [14]. Thus, how LSECs change during the model of liver fibrosis induced by schistosoma infection remains unknown.

In this study, a mouse model of *Schistosoma japonicum* infection was used. LSECs were isolated using gradient density centrifugation from the model mice. The changes in LSEC phenotype and function were detected by flow cytometry. The ultrastructure of LSECs was observed by scanning and transmission electron microscopes. The changes in EMT and fibrosis in LSECs and liver tissues were detected by qPCR and Western blot. This study sought to investigate the changes in LSECs in the infected model of schistosoma and the effect on liver fibrosis. These findings will be helpful to identify an effective strategy to treat liver fibrosis induced by *S. japonicum* infection.

2. Materials and Methods

2.1. Ethics Statement

Our experiments involving C57BL/6 mice were performed according to China's Laboratory of Animal Welfare and Ethics Committee (LAWEC). The LAWEC Committee of the National Institute of Parasitic Diseases Chinese Center for Disease Control and Prevention approved the protocol (NIPD-2020-10).

2.2. Animals and Parasites

Female specific pathogen-free (SPF) C57BL/6 mice (6–8 weeks old; body weight 20 ± 2 g) were purchased from Shanghai Lingchang Biotechnology Co., Ltd. (Shanghai, China). Mice were housed in an SPF-grade animal room at our institute. The vector-borne tropical ward of our institute provided cerariae. Mice were percutaneously infected with cercariae by shaving the skin of the abdomen.

2.3. Reagents

Dulbecco's phosphate-buffered saline (DPBS) and albumin from bovine serum (BSA) were purchased from Gibco (Grand Island, NY, USA). A Percoll cell separation solution was purchased from GE (Chicago, IL, USA), and collagenase from clostridium histolyticum was purchased from Sigma-Aldrich (St. Louis, MO, USA). Fluorescently conjugated antibodies, including PerCP-Cy5.5 rat anti-mouse CD45, FITC rat anti-mouse CD146, Fixable Viability Stain 575V, PE rat anti-mouse CD32b, Bv421 rat anti-mouse TGF-β and AF647 rat anti-mouse eNOS were purchased from Invitrogen Corporation (Waltham, MA, USA). Mouse CD146 kit Invitrogen™ was provided by Miltenyi Biotec Co.,

Ltd. (Bergisch Gladbach, Germany). Hyper ScriptTMIII RT SuperMix and Universal SYBR qPCR Mix were provided by Enzy Artisan (Shanghai, China).

2.4. Infection and Cell Isolation

A total of 60 C57BL/6 mice were randomly divided into an infected or uninfected group (30 mice/group). Mice in the infected group were anesthetized via an intraperitoneal injection of 1% sodium pentobarbital. Mice with shaved abdomen skin were percutaneously infected with 20 ± 1 cercariae. The uninfected group did not receive any treatment.

After anesthetizing the mice, 10 infected and 10 uninfected mice were sacrificed at 20, 28, and 42 days post infection, respectively. Mice were sterilized with 75% alcohol, secured to the board, and perfused with $1\times$ Dulbecco's phosphate-buffered saline (DPBS) to remove red blood cells. After perfusion, the liver was cut into fragments. Collagenase dissociated liver tissue into a single-cell suspension. Cells were separated by differential gradient centrifugation with 25%/50% Percoll solution [15]. The supernatant and lipid layer was discarded, and the cells were washed twice with DPBS. The red cells were lysed using BD Pharm Lyse™ lysing solution (Becton Dickinson and Company, Franklin Lakes, NJ, USA) to obtain hepatic non-parenchymal cells.

The concentrations of the hepatic non-parenchymal cells were adjusted to 1×10^7 cells. We used an LSEC Isolation Kit (MiltenyiBiotec, Auburn, CA, USA) to isolate LSECs. The cellular suspension was centrifuged, and the supernatant was completely removed. The cell pellet was resuspended in 90 μL of buffer (phosphate-buffered saline, pH 7.2; 0.025% bovine serum albumin [BSA]; and 0.1 mM ethylenediaminetetraacetic acid [EDTA]). Next, 10 μL CD146 microbeads per 10^7 total cells were added, and the mixture was incubated at 4 °C for 15 min. The cells were washed by adding 1 mL buffer and centrifuged at $300\times g$ for 10 min, and the supernatant was completely removed using a pipette. The cell precipitates were resuspended with 500 μL buffer. The magnetic rack was placed, and the magnetic sorting column was placed in the magnetic field. Next, 500 μL buffer was added to wash the column, and the cell suspension was passed through the column. The column was washed with buffer three times. The sorting column was removed from the magnetic rack and placed on the appropriate collection tube. Then, 1 mL buffer was added to the column and the magnetic-labeled LSECs were flushed out through the plunger.

2.5. Flow Cytometry

The hepatic non-parenchymal cells were adjusted to 1×10^6/mL using fluorescence-activated cell sorting (FACS) buffer (2% BSA and 2 mM EDTA in DPBS). The following antibodies were used in our experiments: Fixable Viability Stain 575V, CD45-Percp-cy5.5, CD146-FITC, CD32b-PE, and TGF-β Bv421. LSECs were defined as CD45$^-$ CD146$^+$.

Each 1 mL cell suspension was mixed with 1 μL Fixable Viability Stain 575V, incubated in the dark at room temperature for 15 min, and the cells were washed twice. Cells were stained with different combinations of antibodies for 30 min at room temperature (24–26 °C) in the dark, and washed with FACS buffer once. The fixation and permeabilization solution was added, incubated in the dark at 4 °C for 20 min, and the cells were washed twice. The cells were stained with intracellular anti-mouse TGF-β diluted antibody (antibody was diluted with $1 \times$ Perm/Wash buffer) at 4 °C for 30 min, and washed twice. The cells were fixed in 1% paraformaldehyde at 4 °C for 20 min, and the cells were washed once. Next, 200 μL staining buffer was added to each sample, and the cells were suspended for detection. All experiments were carried out using a BD FACS Verse flow cytometer (BD Biosciences). Data were analyzed with FlowJo 10 software (TreeStar Inc., Ashland, OR, USA). The proportions of LSECs, CD32b$^+$ LSECs, TGF-β$^+$ LSECs, and eNOS expression were measured.

2.6. Electron Microscopy

Scanning electron microscope (SEM): after removing the red blood cells from the liver, an electron microscope fixation solution containing 2.5% glutaraldehyde was perfused into the mouse liver. After the liver had hardened, the edge of the right liver lobe was cut to approximately 2 mm × 2 mm and soaked into the electron microscope fixation solution for 2 h. The fixed samples were rinsed three times with phosphate buffer (0.1M) (PB, PH 7.4). The samples were fixed with 1% osmium acid at room temperature for 1–2 h avoiding light, and rinsed with PB three times. The samples were dehydrated, dried, and sprayed with gold for 30 s. The LSEC window pores were observed with SEM and photographed.

Transmission electron microscope (TEM): TEM samples were dehydrated twice with 30–100% alcohol and 100% acetone. The samples were embedded and polymerized at 60 °C for 48 h to prepare 60–80 μm ultra-thin slices. The samples were stained to avoid light with a 2% uranium acetate saturated alcohol solution for 8 min and to prevent carbon dioxide with a lead citrate solution for 8 min. After cleaning and drying, the samples were observed and photographed with a transmission electron microscope to evaluate the changes in the LSEC basement membrane.

2.7. Reverse Transcription Quantitative PCR

The total RNA from LSECs and mouse livers were extracted using TRIzol. Complementary DNA (cDNA) was synthesized using 1 μg total RNA with a Prime Script RT Master Mix (Takara, Shiga). RT-qPCR was used to determine the level of gene expression, including those in the liver samples (E-cadherin, N-cadherin, fibronectin, laminin, vimentin, α-SMA, collagen I, III, and IV) and LSEC samples (E-cadherin, VE-cadherin, Zonula occludens1, fibronectin, and α-SMA) using Fast SYBR Green master Mix (Bio-Rad, Hercules, CA, USA). The qPCR reaction system included 10 μL 2 × S6 Universal SYBR qPCR Mix, 1 μL upstream primers, 1 μL downstream primers, 3 μL cDNA, and 5 μL ddH$_2$O. The primers used in this study are listed in Table 1, which were synthesized by Enzy Artisan Co., LTD (Shanghai, China). The qPCR reaction conditions were as follows: 95 °C for 30 s; 95 °C for 5 s, 60 °C for 30 s, over 38 cycles. After the circulating value (Ct) was obtained, the relative expression of the target gene was evaluated using $2^{-\Delta\Delta Ct}$.

Table 1. Primer sequences for qPCR.

Gene Name	Primer Sequence (5′→3′)
GAPDH	F:CATCACTGCCACCCAGAAGACTG R:ATGCCATGAGCTTCCCGTTCAG
E-cadherin	F:GGTCATCAGTGTGCTCACCTCT R:GCTGTTGTGCTCAAGCCTTCAC
VE-cadherin	F:GAACGAGGACAGCAACTTCACC R:GTTAGCGTGCTGGTTCCAGTCA
Zonula occludens1 (ZO1)	F:GTTGGTACGGTGCCCTGAAAGA R:GCTGACAGGTAGGACAGACGAT
Fibronectin (FN)	F:GGTCCTCTCCTTCCATCTCCTTAC R:GGACCCCTGAGCATCTTGAGTG

2.8. Western Blot

The mouse liver tissues were collected and lysed using a radioimmunoprecipitation assay (RIPA) lysis buffer (Shanghai Epizyme Biomedical Technology Co., Ltd., Shanghai, China) supplemented with a protease inhibitor cocktail and EDTA (Beyotime Biotechnology, Shanghai, China). The lysates were centrifuged at 12,000× *g* for 10 min. The protein concentrations were detected using the BCA method. After boiling at 100 °C for 10 min, the samples were loaded into wells, and SDS-polyacrylamide gel electrophoresis was performed at 90 V for approximately 1 h. The protein strips were transferred to polyvinylidene fluoride (PVDF) membranes. After blocking, the membranes were incubated sequentially with primary and secondary antibodies. Anti-GAPDH (5174S, Cell Signaling Technology, CST, Danvers, MA, USA), anti-a-SMA (19245S, CST), and anti-Collagen I a1 (bs-7158R, Bioss) antibodies were used as the primary antibodies. A horseradish peroxidase-conjugated anti-mouse IgG antibody (7076S, CST) was used as the secondary antibody. Immunoreactive bands were visualized on digital images captured with a ChemiDoc MP Imaging

System (Bio-Rad). The band intensities were quantified using Image J software (NIH, Bethesda, MD, USA).

2.9. Statistical Analysis

Data analysis was performed using GraphPad Prism Version 9.0.0 and SPSS 20.0 (IBM Corp., Armonk, NY, USA). Differences between the groups were assessed using a nonparametric one-way analysis of variance. Data were presented as the mean $\pm$ standard deviation. $p < 0.05$ indicated a significant difference.

3. Results

3.1. The Changes in the Proportion of LSECs in Mice Infected with S. japonicum

The flow cytometry results showed that the percentage of LSECs (CD45-CD146$^+$) in hepatic non-parenchymal cells was (28.70 $\pm$ 6.41)%, (9.43 $\pm$ 4.88)%, and (2.18 $\pm$ 0.49)% at 20, 28, and 42 days, respectively following *S. japonicum* infection. Compared with the uninfected group (50.40 $\pm$ 1.68)%, the proportion of LSECs decreased significantly ($p < 0.01$) (Figure 1A,D). After infection for 20, 28, and 42 days, the proportion of CD32b$^+$ LSECs was (95.80 $\pm$ 0.28)%, (89.22 $\pm$ 4.03)%, and (74.82 $\pm$ 5.06)%, which was lower than that of the uninfected group (97.27 $\pm$ 0.58)% (Figure 1B,E). The TGF-β^+ LSEC population was (80.20 $\pm$ 1.78)%, (88.37 $\pm$ 1.49)%, and (81.83 $\pm$ 3.55)%, which was higher than that of the uninfected group (73.37 $\pm$ 3.44)% (Figure 1C,F). The results indicated that the proportion of differentiated LSECs (CD32b$^+$ LSECs) decreased, whereas the proportion of dedifferentiated LSECs (TGF-β^+ LSECs) significantly increased after infection. The expression of eNOS in LSECs decreased significantly post infection during the 20th to 42th day post infection. This indicated that the secretion of NO in LSECs was decreased, and the function of LSECs was damaged.

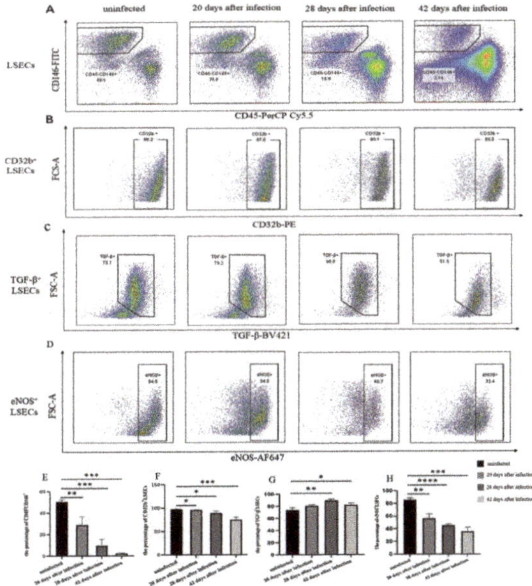

Figure 1. Changes in the LSEC phenotype after being infected with *S. japonicum*. (**A**) The population of LSECs in hepatic non-parenchymal cells after infection. (**B**) The population of CD32b$^+$ LSECs in hepatic non-parenchymal cells after infection. (**C**) The population of TGF-β^+ LSECs in hepatic non-parenchymal cells after infection. (**D**) The population of eNOS$^+$ LSECs in the hepatic non-parenchymal cells after infection. (**E**) Histogram showing the proportion of LSECs (CD45$^-$CD146$^+$). (**F**) Histogram of the proportion of CD32b$^+$ LSECs. (**G**) Histogram of the proportion of TGF-β^+ LSECs. (**H**) Histogram of the proportion of eNOS$^+$ LSECs. * $p < 0.05$; ** $p < 0.01$; *** $p < 0.001$; **** $p < 0.0001$.

3.2. Changes in the Number of Fenestrations of LSEC after Infection

The SEM results showed that the LSECs of the uninfected mice displayed multiple fenestrations, which dispersed and connected to form a sieve plate. The number of fenestrations in LSEC decreased significantly post infection. In contrast, the number of LSECs without fenestration increased continuously (Figure 2A–D). The TEM results revealed that the number of microvilli in the disc space were reduced after infection for 42 days. A large amount of collagen was deposited in the hepatic disc space in the liver of infected mice. In the uninfected group, LSECs in the liver of the mice had obvious fenestration and did not have a basement membrane. After being infected for 42 days, fenestration disappeared, and the basement membrane appeared under the LSECs (Figure 2E–J).

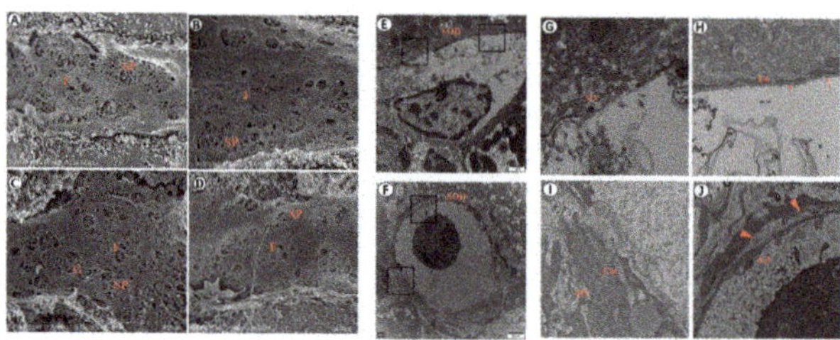

Figure 2. The structure of LSEC fenestration was altered in the liver of mice following infection. (**A**) The number of LSEC fenestration in the uninfected group. (**B**) The number of LSEC fenestration for 20 days after infection. (**C**) The number of LSEC fenestrations for 28 days post infection. (**D**) The number of LSEC fenestrations for 42 days post infection. (**E**) The structure of the LSEC basement membrane in the uninfected group. (**F**) The structure of the LSEC basement membrane for 42 days post infection. (**G**) The number of microvilli in the hepatic disc space of the uninfected mice (TEM magnification). (**H**) The structure of the LSEC basement membrane in the uninfected group (TEM magnification). (**I**) The number of microvilli and collagen in the hepatic disc space for 42 days post infection (TEM magnification). (**J**) The LSEC basement membrane structure for 42 days post infection (TEM magnification). SP, sieve plate; F, fenestration; G, gap; Ed, sinusoidal endothelial cells; SOD, space of Disse; MV, microvilli; Col, collagen; small arrow, fenestration; big arrow, basement membrane.

3.3. The Changes of EMT in LSECs after Infection with S. japonicum

The qPCR results showed that after infection for 20 days, the level of E-cadherin mRNA expression in the LSECs was significantly lower than that in the uninfected group ($p < 0.05$) (Figure 3A). However, the levels of VE-cadherin, ZO1, fibronectin, and α-SMA mRNA expression in LSECs did not differ from that of the uninfected group. Following infection for 28 and 42 days, the level of E-cadherin, VE-cadherin, and ZO1 mRNA expression in LSECs decreased (Figure 3A–C). In contrast, the levels of fibronectin and α-SMA mRNA expression increased significantly (Figure 3D–E). These results indicated that LSECs began EMT changes during the early stages after infection, and the LSECs transformed into fibroblasts during the middle and late stages, producing large amounts of α-SMA.

3.4. EMT and Liver Fibrosis in the Liver after Infection with S. japonicum

The qPCR results showed that the levels of mRNA markers associated with EMT (vimentin) and liver fibrosis (Collagen I, III, and IV) in the liver at 20 days post infection were significantly higher than those in the uninfected group (Figure 4E,G–I). After infection for 28 and 42 days, the level of E-cadherin mRNA decreased (Figure 4A), whereas the level of marker mRNA associated with EMT (N-cadherin, fibronectin, laminin, and vimentin) were increased significantly (Figure 4B–E). The levels of marker mRNA associated with liver fibrosis, including α-SMA, collagen I, III, and IV, were increased at 28 and 42 days post infection (Figure 4F–I). The Western blot results showed that the protein levels of α-SMA and collagen I were increased consistently at 28 and 42 days post infection (Figure 4J–L). These results indicated that at 20 days post infection, the liver began to undergo moderate changes in fibrosis and EMT. After infection for 28 and 42 days, the changes in fibrosis and EMT in the liver became more significant.

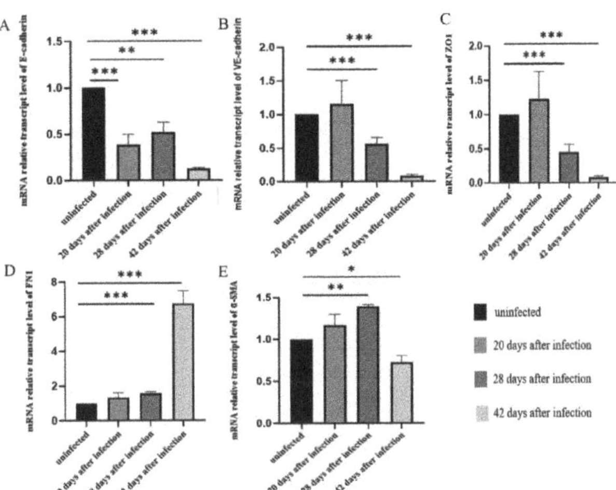

Figure 3. LSEC EMT after *S. japonicum* infection. (**A**) The level of E-cadherin mRNA expression in LSECs in mice post infection. (**B**) The level of VE-cadherin mRNA expression in LSECs in mice post infection. (**C**) ZO1 expression of LSECs in mice post infection. (**D**) FN1 expression in LSECs in mice post infection. (**E**) α-SMA expression in LSECs in the mice post infection. * $p < 0.05$; ** $p < 0.01$; *** $p < 0.001$.

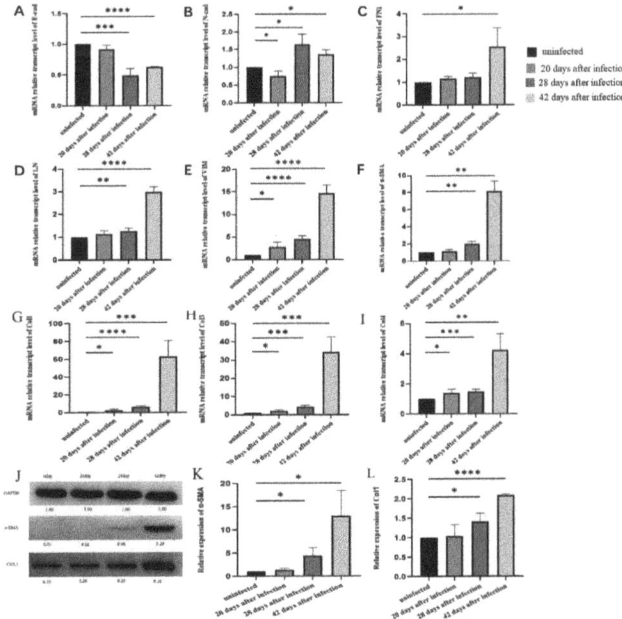

Figure 4. Fibrosis and EMT changes in the liver following infection with *S. japonicum*. (**A–E**) qPCR results of E-cadherin, N-cadherin, FN1, LN, and vimentin expression in the liver after infection. (**F–I**) qPCR results of α-SMA, collagen I, II, and III expression in the liver after infection. (**J–L**) Western blot results of α-SMA and collagen I expression in the liver after infection. * $p < 0.05$; ** $p < 0.01$; *** $p < 0.001$; **** $p < 0.0001$.

4. Discussion

Schistosomiasis is a globally distributed neglected tropical disease, which primarily occurs in tropical and subtropical regions [16]. Worldwide, approximately 236 million people are infected with schistosomes, 90% of which are located in sub-Saharan Africa, and results in approximately 300,000 deaths per year [17,18]. In China, according to the national schistosomiasis epidemic bulletin in 2020, although schistosomiasis has a low prevalence, there remains a large number of patients with schistosomiasis liver fibrosis [2]. At present, there is no effective treatment for liver fibrosis induced by schistosoma infection. Therefore, elucidating the regulatory mechanism of schistosomiasis liver fibrosis is essential to providing potential targets for the treatment of liver fibrosis.

After being infected with *S. japonicum*, the eggs are deposited in the host liver. Several antigens from the schistosomula and soluble eggs that are released induce both immune and inflammatory responses [19,20]. Some cells (e.g., hepatocytes, hepatic stellate cells, hepatic sinusoidal endothelial cells, and bile duct epithelial cells) transform into myofibroblasts (MFB), releasing large amounts of extracellular matrix (ECM) [21]. The mass transformation to MFB represents a key link in the development of hepatic fibrosis. A large amount of ECM deposits in the liver tissue leads to the formation of schistosomiasis liver fibrosis and seriously affects human health [22].

LSECs are at the highest proportion in mouse livers, accounting for approximately 70% of liver non-parenchymal cells [23]. Moreover, LSECs play an essential role in maintaining the balance between liver regeneration and fibrosis [10]. LSECs have differentiation and dedifferentiation phenotypes. The differentiated LSECs have a normal function, which are rich in fenestration and lack a complete basement membrane. The permeability of hepatic sinuses depends on this unique structure, which promotes the exchange of nutrients and gases between cells. Nitric oxide synthase (eNOS) activity is high in differentiated LSECs, which produce and release NO, and maintain the resting state of HSCs [8]. In chronic liver injury, the fenestration diameter of LSECs decreased, and a complete basement membrane gradually formed. This results in changes to LSEC dedifferentiation [12]. During the early stage of non-alcoholic liver injury, chronic hepatitis, and other models, LSECs exhibit changes in dedifferentiation [13,24].

The key feature of MFB activation is epithelial–mesenchymal transition (EMT) [21,25–27]. During the change in EMT, epithelium acquires mesenchymal properties, which plays an important role in tissue repair, inflammation, fibrosis, and other processes [28,29]. EMT is an important source of myofibroblasts [10]. Growing evidence shows that when EMT is dominant in tissues, liver tissues progress toward fibrosis [27]. Dedifferentiated LSECs transform into myofibroblast-like cells through EMT and secrete a large amount of fibronectin and α-SMA to promote fibrosis formation [30–32]. Therefore, maintaining LSECs with a differentiated phenotype is an effective strategy for reversing hepatic fibrosis.

After infection, LSECs developed a dedifferentiated phenotype and changed into mesenchymal cells. CD32b is an essential indicator of differentiated LSEC, and high levels of TGF-β were detected in dedifferentiated LSEC [12,33]. Following infection for 20, 28, and 42 days, the proportion of total LSECs (CD45$^-$CD146$^+$) and differentiation phenotype CD32b$^+$ LSECs in infected mice decreased continuously, whereas the dedifferentiation phenotype (TGF-β^+ LSECs) increased significantly. Following infection, the level of eNOS protein expression in LSECs decreased significantly, indicating that the amount of NO secreted by LSECs had decreased significantly. This finding suggested that the ability of LSECs to maintain the resting state of HSCs was reduced. Moreover, the number of fenestrations in LSECs decreased and the basement membrane formed. This finding suggested that LSECs converted to a dedifferentiated phenotype in the middle and late stages of infection. EMT is a dynamic pathological process. Some of the dedifferentiated LSECs underwent further EMT changes and transformed into mesenchymal cells, producing a large amount of ECM. The level of epithelial marker mRNA, including E-cadherin, VE-cadherin, and Zo1, decreased, while mesenchymal markers (FN1 and α-SMA) increased during the early stages. Additionally, they changed more significantly during the middle and late stages of infection. These results indicated that the dedifferentiation and EMT in LSECs began during the early stages of infection (20 days post infection). With the development of infection, the extent of pathological changes aggravated in the LSECs.

Pathological changes in LSECs could further promote the progression of liver fibrosis induced by *S. japonicum* infection. On the one hand, dedifferentiated LSECs could induce HSC activation. Dedifferentiated LSECs reduced NO secretion but greatly increased TGF-β secretion post infection. NO benefits the maintenance of HSC quiescence [11,34], whereas TGF-β is a classical cytokine that activates HSCs [35]. Activated HSCs were the primary source of MFB [36,37]. Therefore, dedifferentiated LSECs had transformed into a pro-inflammatory phenotype, which could indirectly promote liver fibrosis by activating HSCs. On the other hand, dedifferentiated LSECs could directly undergo EMT changes, transform into MFB, produce a large number of ECM, and promote hepatic fibrosis [38,39]. Maintaining the differentiated phenotype of LSEC is an effective treatment for hepatic fibrosis. Some studies have shown that the administration of statin treatment or actin depolymerization to maintain the LSEC phenotype could decrease portal pressure and improve NASH features in an early NASH model [33,40]. Long noncoding RNAs can interact with EZH2, and maintain LSEC differentiation through KLF2-eNOS-sGC pathway and alleviate CCl$_4$-induced liver fibrosis [41].

It has been well established that egg deposition in the liver releases soluble egg antigens and recruits lymphocytes, leading to the formation of granuloma and liver fibrosis [42,43]. However, our experiments indicated that egg deposition in the liver was not the only cause of liver fibrosis. LSEC injury during the early stage of infection also promoted the progression of liver fibrosis. It was established that *S. japonicum* lay eggs about 24 days post infection. The pathological changes in LSECs occurred earlier than schistosome oviposition. Pathological lesions appeared in LSECs on the 20th day or earlier after infection. Pathological injury to LSECs could also induce EMT and fibrosis changes in the liver tissue. The qPCR and Western blot results showed that markers of EMT and fibrosis in the liver changed during the early stage (20th day post infection) and aggravated in the middle and late stages of infection. The pathological changes in the liver were consistent with that of LSECs. Thus, histological changes in LSECs induced by *S. japonicum* infection are also important factors for promoting liver fibrosis.

However, our study only simultaneously observed EMT and fibrosis changes in LSEC and liver tissue in schistosomiasis and did not determine whether EMT changes occurred earlier than liver fibrosis. It also remains unknown whether EMT changes cause liver fibrosis. Therefore, further experiments are needed to explore the role of EMT in liver fibrosis.

5. Conclusions

In the present study, we identified changes in the pathology and functional impairment of LSEC following an infection with *S. japonicum*. These changes predated schistosome oviposition. The changes in LSEC were consistent with the process of liver fibrosis induced by *S. japonicum* infection. This indicates that LSECs are involved in the regulation of schistosomiasis liver fibrosis. Therefore, using drugs or Lnc RNA to target the pathological changes in LSECs represents a promising strategy that can alleviate or reverse schistosomiasis liver fibrosis.

Author Contributions: Conceptualization and design, Y.H., J.C. and T.J.; supervision, Y.H. and J.C.; methodology, formal analysis, writing—original draft, T.J., X.W. and H.Z.; writing—review and editing, J.C. and Y.H. All authors have read and agreed to the published version of the manuscript.

Funding: This work was funded by the Shanghai Natural Science Foundation (grant number: 19ZR1462600 to Y.H.), and the National Natural Science Foundation of China (grant numbers: 81971969 and 82272369 to J.C.).

Institutional Review Board Statement: The animal study protocol was approved by the Laboratory of Animal Welfare and Ethics Committee (LAWEC), National Institute of Parasitic Diseases, Chinese Center for Disease Control and Prevention, China (Permit Number: NIPD-2020-10).

Data Availability Statement: Data are contained within the article.

Acknowledgments: We thank our colleagues in the Animal House who raised laboratory animals. We also thank our colleagues in the Instrument Room Center for their maintenance and management of the instruments.

Conflicts of Interest: The authors have no conflict of interest to declare. The funders had no role in the design of the study; in the collection, analyses, or interpretation of data; in the writing of the manuscript; or in the decision to publish the results.

References

1. LoVerde, P.T. Schistosomiasis. *Adv. Exp. Med. Biol.* **2019**, *1154*, 45–70. [PubMed]
2. Zhang, L.J.; Xu, Z.M.; Yang, F.; Dang, H.; Li, Y.L.; Lv, S.; Chao, C.L.; Xu, J.; Li, S.Z.; Zhou, X.L. Endemic status of schistosomiasis in People's Republic of China in 2020. *Chin. J. Schistosomiasis Control* **2021**, *33*, 225–233.
3. Wang, W.; Bergquist, R.; King, C.H.; Yang, K. Elimination of schistosomiasis in China: Current status and future prospects. *PLoS Negl. Trop. Dis.* **2021**, *15*, e0009578. [CrossRef] [PubMed]
4. Gong, Y.F.; Zhu, L.Q.; Li, Y.L.; Zhang, L.J.; Xue, J.B.; Xia, S.; Lv, S.; Xu, J.; Li, S.Z. Identification of the high-risk area for schistosomiasis transmission in China based on information value and machine learning: A newly data-driven modeling attempt. *Infect. Dis. Poverty* **2021**, *10*, 88. [CrossRef] [PubMed]
5. Lo, N.C.; Bezerra, F.S.M.; Colley, D.G.; Fleming, F.M.; Homeida, M.; Kabatereine, N.; Kabole, F.M.; King, C.H.; Mafe, M.A.; Midzi, N.; et al. Review of 2022 WHO guidelines on the control and elimination of schistosomiasis. *Lancet. Infect. Dis.* **2022**, *22*, e327–e335. [CrossRef]
6. Qin, Z.Q.; Xu, J.; Feng, T.; Lv, S.; Qian, Y.J.; Zhang, L.J.; Li, Y.L.; Lv, C.; Bergquist, R.; Li, S.Z.; et al. Field Evaluation of a Loop-Mediated Isothermal Amplification (LAMP) Platform for the Detection of Schistosoma japonicum Infection in Oncomelania hupensis Snails. *Trop. Med. Infect. Dis.* **2018**, *3*, 124. [CrossRef]
7. Kisseleva, T.; Brenner, D. Molecular and cellular mechanisms of liver fibrosis and its regression. *Nat. Rev. Gastroenterol. Hepatol.* **2021**, *18*, 151–166. [CrossRef]
8. Shetty, S.; Lalor, P.F.; Adams, D.H. Liver sinusoidal endothelial cells—Gatekeepers of hepatic immunity. *Nat. Rev. Gastroenterol. Hepatol.* **2018**, *15*, 555–567. [CrossRef]
9. Poisson, J.; Lemoinne, S.; Boulanger, C.; Durand, F.; Moreau, R.; Valla, D.; Rautou, P.E. Liver sinusoidal endothelial cells: Physiology and role in liver diseases. *J. Hepatol.* **2017**, *66*, 212–227. [CrossRef]
10. Lafoz, E.; Ruart, M.; Anton, A.; Oncins, A.; Hernández-Gea, V. The endothelium as a driver of liver fibrosis and regeneration. *Cells* **2020**, *9*, 929. [CrossRef]
11. Deleve, L.D.; Wang, X.; Guo, Y. Sinusoidal endothelial cells prevent rat stellate cell activation and promote reversion to quiescence. *Hepatology* **2008**, *48*, 920–930. [CrossRef] [PubMed]
12. Ma, H.; Liu, X.; Zhang, M.; Niu, J.Q. Liver sinusoidal endothelial cells are implicated in multiple fibrotic mechanisms. *Mol. Biol. Rep.* **2021**, *48*, 2803–2815. [CrossRef] [PubMed]
13. Hammoutene, A.; Rautou, P.E. Role of liver sinusoidal endothelial cells in non-alcoholic fatty liver disease. *J. Hepatol.* **2019**, *70*, 1278–1291. [CrossRef] [PubMed]
14. Ribera, J.; Pauta, M.; Melgar-Lesmes, P.; Córdoba, B.; Bosch, A.; Calvo, M.; Rodrigo-Torres, D.; Sancho-Bru, P.; Mira, A.; Jiménez, W.; et al. A small population of liver endothelial cells undergoes endothelial-to-mesenchymal transition in response to chronic liver injury. *Am. J. Physiol. Gastrointest. Liver Physiol.* **2017**, *313*, G492–G504. [CrossRef]
15. Kochan, K.; Kus, E.; Filipek, A.; Szafrańska, K.; Chlopicki, S.; Baranska, M. Label-free spectroscopic characterization of liver liver sinusoidal endothelial cells (LSECs) isolated from the murine liver. *Analyst* **2017**, *142*, 1308–1319. [CrossRef]
16. Deol, A.K.; Fleming, F.M.; Calvo-Urbano, B.; Walker, M.; Bucumi, V.; Gnandou, I.; Tukahebwa, E.M.; Jemu, S.; Mwingira, U.; Alkohlani, A.; et al. Schistosomiasis—Assessing progress toward the 2020 and 2025 global goals. *N. Engl. J. Med.* **2019**, *381*, 2519–2528. [CrossRef]
17. Carbonell, C.; Rodríguez-Alonso, B.; López-Bernús, A.; Almeida, H.; Galindo-Pérez, I.; Velasco-Tirado, V.; Marcos, M.; Pardo-Lledías, J.; Belhassen-García, M. Clinical spectrum of schistosomiasis: An update. *J. Clin. Med.* **2021**, *10*, 5521. [CrossRef]
18. McManus, D.P. The search for a schistosomiasis vaccine: Australia's contribution. *Vaccines* **2021**, *9*, 872. [CrossRef]
19. McManus, D.P.; Dunne, D.W.; Sacko, M.; Utzinger, J.; Vennervald, B.J.; Zhou, X.N. Schistosomiasis. *Nat. Rev. Dis. Prim.* **2018**, *4*, 13. [CrossRef]
20. Zhu, J.F.; Xu, Z.P.; Chen, X.J.; Zhou, S.; Zhang, W.W.; Chi, Y.; Li, W.; Song, X.; Liu, F.; Su, C. Parasitic antigens alter macrophage polarization during *Schistosoma japonicum* infection in mice. *Parasit Vectors* **2014**, *7*, 122. [CrossRef]
21. Chen, Y.; Fan, Y.; Guo, D.Y.; Xu, B.; Shi, X.Y.; Li, J.T.; Duan, L.F. Study on the relationship between hepatic fibrosis and epithelial-mesenchymal transition in intrahepatic cells. *Biomed. Pharmacother.* **2020**, *129*, 110413. [CrossRef]

22. Villesen, I.F.; Daniels, S.J.; Leeming, D.J.; Karsdal, M.A.; Nielsen, M.J. Review article: The signalling and functional role of the extracellular matrix in the development of liver fibrosis. Aliment. *Pharmacol. Ther.* **2020**, *52*, 85–97. [CrossRef] [PubMed]
23. Yuan, B.; Zhai, S.M.; Huang, H.F.; Zeng, Z. A balance mediated by liver sinusoidal endothelial cells on liver regeneration and fibrosis. *Chin. J. Hepatobiliary Surg.* **2021**, *27*, 545–548.
24. Hu, J.; Srivastava, K.; Wieland, M.; Runge, A.; Mogler, C.; Besemfelder, E.; Terhardt, D.; Vogel, M.J.; Cao, L.J.; Korn, C.; et al. Endothelial cell-derived angiopoietin-2 controls liver regeneration as a spatiotemporal rheostat. *Science* **2014**, *343*, 416–419. [CrossRef] [PubMed]
25. Zhang, Q.D.; Xu, M.Y.; Cai, X.B.; Qu, Y.; Li, Z.H.; Lu, L.G. Myofibroblastic transformation of rat hepatic stellate cells: The role of Notch signaling and epithelial-mesenchymal transition regulation. *Eur. Rev. Med. Pharmacol. Sci.* **2015**, *19*, 4130–4138. [PubMed]
26. Ikegami, T.; Zhang, Y.; Matsuzaki, Y. Liver fibrosis: Possible involvement of EMT. *Cells Tissues Organs* **2007**, *185*, 213–221. [CrossRef] [PubMed]
27. Cicchini, C.; Amicone, L.; Alonzi, T.; Marchetti, A.; Mancone, C.; Tripodi, M. Molecular mechanisms controlling the phenotype and the EMT/MET dynamics of hepatocyte. *Liver Int.* **2015**, *35*, 302–310. [CrossRef]
28. Jayachandran, J.; Srinivasan, H.; Mani, K.P. Molecular mechanism involved in epithelial to mesenchymal transition. *Arch. Biochem. Biophys.* **2021**, *710*, 108984. [CrossRef]
29. Choi, S.S.; Diehl, A.M. Epithelial-to-mesenchymal transitions in the liver. *Hepatology* **2009**, *50*, 2007–2013. [CrossRef]
30. Hammoutene, A.; Biquard, L.; Lasselin, J.; Kheloufi, M.; Tanguy, M.; Vion, A.C.; Mérian, J.; Colnot, N.; Loyer, X.; Tedgui, A.; et al. A defect in endothelial autophagy occurs in patients with non-alcoholic steatohepatitis and promotes inflammation and fibrosis. *J. Hepatol.* **2020**, *72*, 528–538. [CrossRef]
31. Wilkinson, A.L.; Qurashi, M.; Shetty, S. The role of sinusoidal endothelial cells in the axis of inflammation and cancer within the liver. *Front. Physiol.* **2020**, *11*, 990. [CrossRef] [PubMed]
32. Yu, Y.P.; Chen, Z.Y.; Ye, L. Research progress of molecular mechanism in epithelial-mesenchymal transformation (EMT) in hepatic fibrosis. *J. Med. Theory Pract.* **2019**, *18*, 2886–2888.
33. Bravo, M.; Raurell, I.; Hide, D.; Fernández-Iglesias, A.; Gil, M.; Barberá, A.; Salcedo, M.T.; Augustin, S.; Genescà, J.; Martell, M. Restoration of liver sinusoidal cell phenotypes by statins improves portal hypertension and histology in rats with NASH. *Sci. Rep.* **2019**, *9*, 20183. [CrossRef] [PubMed]
34. Ruart, M.; Chavarria, L.; Campreciós, G.; Suárez-Herrera, N.; Montironi, C.; Guixé-Muntet, S.; Bosch, J.; Friedman, S.L.; Garcia-Pagán, J.C.; Hernández-Gea, V. Impaired endothelial autophagy promotes liver fibrosis by aggravating the oxidative stress response during acute liver injury. *J. Hepatol.* **2019**, *70*, 458–469. [CrossRef]
35. Su, T.H.; Kao, J.H.; Liu, C.J. Molecular mechanism and treatment of viral hepatitis-related liver fibrosis. *Int. J. Mol. Sci.* **2014**, *15*, 10578–10604. [CrossRef]
36. Carson, J.P.; Ramm, G.A.; Robinson, M.W.; McManus, D.P.; Gobert, G.N. Schistosome-induced fibrotic disease: The role of hepatic stellate cells. *Trends. Parasitol.* **2018**, *4*, 524–540. [CrossRef]
37. Berumen, J.; Baglieri, J.; Kisseleva, T.; Mekeel, K. Liver fibrosis: Pathophysiology and clinical implications. *Wiley Interdiscip. Rev. Syst. Biol. Med.* **2021**, *13*, e1499. [CrossRef]
38. Wang, Y.; Liu, Y. Gut-liver-axis: Barrier function of liver sinusoidal endothelial cell. *J. Gastroenterol. Hepatol.* **2021**, *36*, 2706–2714. [CrossRef]
39. Wu, X.; Shu, L.; Zhang, Z.; Li, J.; Zong, J.; Cheong, L.Y.; Ye, D.; Lam, K.S.L.; Song, E.; Wang, C.; et al. Adipocyte fatty acid binding protein promotes the onset and progression of liver fibrosis via mediating the crosstalk between liver sinusoidal endothelial cells and hepatic stellate cells. *Adv. Sci.* **2021**, *8*, e2003721. [CrossRef]
40. Di Martino, J.; Mascalchi, P.; Legros, P.; Lacomme, S.; Gontier, E.; Bioulac-Sage, P.; Balabaud, C.; Moreau, V.; Saltel, F. Actin depolymerization in dedifferentiated liver sinusoidal endothelial cells promotes fenestrae re-formation. *Hepatol. Commun.* **2018**, *3*, 213–219. [CrossRef]
41. Chen, T.; Shi, Z.; Zhao, Y.; Meng, X.; Zhao, S.; Zheng, L.; Han, X.; Hu, Z.; Yao, Q.; Lin, H.; et al. LncRNA Airn maintains LSEC differentiation to alleviate liver fibrosis via the KLF2-eNOS-sGC pathway. *BMC Med.* **2022**, *20*, 335. [CrossRef] [PubMed]
42. McManus, D.P.; Bergquist, R.; Cai, P.; Ranasinghe, S.; Tebeje, B.M.; You, H. Schistosomiasis-from immunopathology to vaccines. *Semin. Immunopathol.* **2020**, *42*, 355–371. [CrossRef] [PubMed]
43. Gryseels, B.; Polman, K.; Clerinx, J.; Kestens, L. Human schistosomiasis. *Lancet* **2006**, *368*, 1106–1118. [CrossRef] [PubMed]

Tropical Medicine and Infectious Disease

Article

Accuracy of Three Serological Techniques for the Diagnosis of Imported Schistosomiasis in Real Clinical Practice: Not All in the Same Boat

María Pilar Luzón-García [1], María Isabel Cabeza-Barrera [1], Ana Belén Lozano-Serrano [1], Manuel Jesús Soriano-Pérez [1], Nerea Castillo-Fernández [1], José Vázquez-Villegas [2], Jaime Borrego-Jiménez [1] and Joaquín Salas-Coronas [1,3,*]

[1] Tropical Medicine Unit, Hospital Universitario Poniente, Ctra. de Almerimar 31, 04700 El Ejido, Spain
[2] Tropical Medicine Unit, Distrito Poniente de Almería, 04700 El Ejido, Spain
[3] Department of Nursing, Physiotherapy and Medicine, Faculty of Health Sciences, University of Almería, 04120 La Cañada, Spain
* Correspondence: joaquin.salas.sspa@juntadeandalucia.es

Abstract: Schistosomiasis is a neglected tropical disease despite of being a major public health problem affecting nearly 240 million people in the world. Due to the migratory flow from endemic countries to Western countries, an increasing number of cases is being diagnosed in non-endemic areas, generally in migrants or people visiting these areas. Serology is the recommended method for screening and diagnosis of schistosomiasis in migrants from endemic regions. However, serological techniques have a highly variable sensitivity. The aim of this study was to evaluate retrospectively the sensitivity of three different serological tests used in real clinical practice for the screening and diagnosis of imported schistosomiasis in sub-Saharan migrant patients, using the detection of schistosome eggs in urine, faeces or tissues as the gold standard. We evaluated three different serological techniques in 405 sub-Saharan patients with confirmed schistosomiasis treated between 2004 and 2022: an enzyme-linked immunosorbent assay (ELISA), an indirect haemagglutination assay (IHA) and an immunochromatographic test (ICT). The overall sensitivity values obtained with the different techniques were: 44.4% for IHA, 71.2% for ELISA and 94.7% for ICT, respectively. According to species, ICT showed the highest sensitivity (*S. haematobium*: 94%, *S. mansoni*: 93.3%; and *S. intercalatum/guineensis*: 100%). In conclusion, our study shows that Schistosoma ICT has the best performance in real clinical practice, when compared to ELISA and IHA, in both *S. mansoni* and *S. haematobium* infections.

Keywords: schistosomiasis; *Schistosoma haematobium*; *Schistosoma mansoni*; diagnosis; serology; immunochromatography; migrants

Citation: Luzón-García, M.P.; Cabeza-Barrera, M.I.; Lozano-Serrano, A.B.; Soriano-Pérez, M.J.; Castillo-Fernández, N.; Vázquez-Villegas, J.; Borrego-Jiménez, J.; Salas-Coronas, J. Accuracy of Three Serological Techniques for the Diagnosis of Imported Schistosomiasis in Real Clinical Practice: Not All in the Same Boat. *Trop. Med. Infect. Dis.* **2023**, *8*, 73. https://doi.org/10.3390/tropicalmed8020073

Academic Editor: Vyacheslav Yurchenko

Received: 20 December 2022
Revised: 14 January 2023
Accepted: 17 January 2023
Published: 19 January 2023

1. Introduction

Schistosomiasis is a neglected tropical disease despite of being a major public health problem affecting nearly 240 million people in at least 78 countries, with more than 700 million living in areas at risk of infection [1–3]. Ninety-three per cent of cases occur in sub-Saharan Africa. It is caused by blood flukes of the genus *Schistosoma*. Seven species can infect humans, although most cases are caused by *S. haematobium* and *S. mansoni*, responsible for urogenital and intestinal schistosomiasis, respectively [4]. Human infections with the remaining species are much less frequent and are restricted to their intermediate host distribution [5,6]. The disease can lead to serious complications causing the death of about 300,000 people each year.

The significant migratory flow experienced in recent decades from endemic countries to Western countries has led to an increase in the diagnosis of imported diseases. These diagnoses may be delayed by language barriers, bureaucracy, access to health care, lack of knowledge of the disease amid health care personnel, poor symptomatology or diagnostic difficulties, among other causes [7]. Regarding schistosomiasis, an increasing number of publications present series of cases diagnosed in non-endemic areas [8–13], generally in migrants or people visiting these areas. On the other hand, outbreaks of autochthonous transmission of *S. haematobium* have recently been described in Corsica (France) and Almería (Spain) [14,15].

In 2019, the European Centre for Disease Prevention and Control (ECDC) recommended serological screening for schistosomiasis in migrants from endemic regions staying in Europe for less than 5 years [3].

Although direct visualisation of *Schistosoma* spp. eggs by microscopic examination of urine and faeces is considered the gold standard for diagnosis, it has limitations, such as low sensitivity, especially in non-endemic areas or in cases of acute infection, and the need for experienced personnel [6]. Antigen detection tests, such as circulating anodic antigen (CAA) and circulating cathodic antigen (CCA), and molecular tests (polymerase chain reaction [PCR] or loop-mediated isothermal amplification of nucleic acids [LAMP]) are more sensitive and potentially useful for the diagnosis of schistosomiasis at all stages and for treatment evaluation [5], but their use is often limited to reference centres [16–19].

Serological tests are currently considered the most effective method for the detection of schistosomiasis in low endemicity and low prevalence settings as they are more sensitive compared to traditional parasitological methods, although they cannot differentiate between active or past disease, and in cases of acute infection, antibodies develop within 6–8 weeks after primary infection [17,20]. Numerous serological techniques, such as indirect haemagglutination (IHA), indirect immunofluorescence assay (IFAT), enzyme-linked immunosorbent assay (ELISA) or immunochromatography-based rapid diagnostic tests (ICT), are commercially available. Most use antigens specific to *S. mansoni* at different stages of the life cycle (adult worm, soluble egg, cercarial antigens, etc.) although these proteins are sufficiently similar to diagnose infections by other *Schistosoma* species with high sensitivity [17,18]. One of the drawbacks of serological testing is that the different techniques have a highly variable sensitivity [21–23] and that is why the correct choice of a particular technique is of great importance, especially for screening programmes.

The aim of this study is to evaluate the sensitivity of three different serological techniques used in real clinical practice for the screening and diagnosis of imported schistosomiasis in sub-Saharan migrant patients, using the detection of schistosome eggs in urine, faeces or tissues as the gold standard.

2. Materials and Methods

A retrospective observational study of sub-Saharan migrant patients with parasitologically confirmed schistosomiasis treated at the Tropical Medicine Unit (TMU) of the Poniente University Hospital (El Ejido, Almería, Spain) from September 2005 to July 2022 was conducted. The Poniente area is an administrative area located in Southeast Spain holding a population close to 300,000 inhabitants with migrants accounting for 21% of the population, many of them coming from sub-Saharan countries to work in horticultural greenhouses.

In order to attend to such residents, there is a screening protocol in Primary Care consisting in a series of laboratory tests aimed to detect imported and cosmopolitan diseases. The tests are usually offered the first-time migrants contact the public health care system, no matter the reason they consult for. For sub-Saharan migrants, the screening protocol includes blood count, liver and renal function tests, syphilis, HIV, HBV and HCV serologies, tuberculin skin test and search for stool parasites and urine parasites.

Whenever an imported disease is either suspected or diagnosed, migrant patients are referred to the hospital Tropical Medicine Clinic. The hospital's protocol is ampler and comprises medical history, epidemiological data, complete physical examination and several additional tests: *Strongyloides* and *Schistosoma* serologies, and Knott and/or saponin tests for microfilariae. Chest and abdominal X-rays are routinely performed too. If any other specific disease is suspected (e.g., onchocerciasis, malaria, etc.), further proper diagnostic procedures are performed.

Parasitological diagnosis of schistosomiasis was made by microscopic visualisation of *S. mansoni* and *S. intercalatum/guineensis* eggs in stool samples (Ritchie's technique in three samples collected on alternate days); of *S. haematobium* eggs in urine (10 mL. of a single sample obtained ideally 10–14 h after light physical exercise), or detection of *Schistosoma* spp. eggs in biopsy samples from different tissues, mainly bladder and rectum.

According to organ involvement, we classified schistosomiasis as hepatointestinal (HI) if *S. mansoni* or *S. intercalatum/guineensis* eggs were demonstrated in faeces or *Schistosoma* spp. eggs were seen in rectal or appendicular biopsy. Urogenital schistosomiasis (UG) was considered when *S. haematobium* eggs were visualized in urine or *Schistosoma* spp. eggs were detected in bladder, cervix or testicular biopsy.

Throughout the study period, three different commercial serological tests were used to detect antibodies to *Schistosoma* spp. in the sera of the study patients: (i) an enzyme-linked immunosorbent assay (ELISA), *Schistosoma mansoni* IgG-ELISA (NovaLisaTM. NovaTec Immundiagnostica, Dietzenbach, Germany); (ii) an indirect haemagglutination test (Bilharziose Fumouze IHA. Famouze Diagnostics, Levallois-Perret, France); (iii) an immunochromatographic rapid diagnostic test (Schistosoma ICT IgG-IgM®. LDBIO Diagnostics, Lyon, France) that simultaneously detects IgM and IgG antibodies against *Schistosoma* spp.

For antibody detection, *S. mansoni* IgG-ELISA NovaLisa TM used a soluble worm antigen preparation [22]. Results were interpreted as negative (index < 1), indeterminate (index 1–1.1) and positive (index > 1.1). For indirect haemagglutination, sensitized sheep erythrocytes coated with *S. mansoni* adult worm antigen were used [24] and a titre $\geq$1:160 was considered positive. For the Schistosoma ICT immunochromatographic test, nitrocellulose strips were coated with an antigen purified from crude lysate of adult *S. mansoni* worms. The test was considered positive if both the control and test strips were positive [25].

The use of one technique or another was determined by their availability in our laboratory. From 2005 to 2013 and from 2015 to 2017, serum samples were evaluated by *Schistosoma mansoni* IgG ELISA (NovaLisaTM). During 2014, the technique used was indirect haemagglutination (Famouze Diagnostics, Levallois-Perret, France). As of January 2018, *Schistosoma* spp. serology was performed in our centre using the Schistosoma ICT IgG-IgM® immunochromatographic test (LDBIO Diagnostics).

We defined the sensitivity of the technique as the proportion of patients with a positive test result among those with a parasitologically proven infection. Schistosomiasis was considered confirmed when *Schistosoma* spp. eggs were detected in faeces, urine or tissues.

A descriptive statistical analysis was performed including all patients diagnosed with confirmed schistosomiasis. Quantitative variables were expressed as mean $\pm$ standard deviation or median $\pm$ interquartile range. Qualitative variables were expressed as frequencies and percentages. Data were analysed using the statistical software package SPSS v17.

3. Results

A total of 405 sub-Saharan migrants with confirmed schistosomiasis were included in the study. Table 1 shows the epidemiological, clinical and analytical characteristics of the patients. The majority (93.6%) were men with a mean age of 27 years (11–52). The mean length of stay in Spain was 35.7 months (1–288). The main countries of origin were Mali (n = 199; 49%), Senegal (n = 92; 22.7%) and Mauritania (n = 33; 8.1%). The most frequent reasons for referral to the TMU were macroscopic hematuria (29.1%), abdominal pain (26.4%) and eosinophilia (16.8%). The rest of the patients were referred because of reasons other than schistosomiasis, mainly because of chronic hepatitis B.

Table 1. Epidemiological, clinical characteristics and laboratory results of patients with confirmed schistosomiasis.

Total of Patients	N = 405
Mean age in years (range, standard deviation)	27 (11–52) SD 6.46
Gender (number, %)	
Male	379 (93.6%)
Mean time living in Spain in months (range, standard deviation)	35.7 (1–288) SD 37.94
Country of (number, %)	
Mali	199 (49%)
Senegal	92 (22.7%)
Mauritania	33 (8.1%)
Equatorial Guinea	16 (4%)
Guinea Conakry	15 (3.7%)
Gambia	12 (3%)
Guinea Bissau	11 (2.7%)
Ghana	11 (2.7%)
Burkina Faso	7 (1.7%)
Ivory Coast	5 (1.2%)
Nigeria	3 (0.7%)
Sierra Leona	1 (0.2%)
Main reason for referral (number, %)	
Macroscopic hematuria	118 (29.1%)
Abdominal pain	107 (26.4%)
Eosinophilia	68 (16.8%)
Microscopic hematuria	12 (3%)
Anemia	4 (1%)
Laboratory tests results (mean, standard deviation)	
Haemoglobin (gr/dL)	14.7 (1.57)
Total eosinophils (Eo/µL)	640 (689.95)
Platelets (Plt/µL)	224 × 103 (68.57)
IgE (IU/L)	2725 (4102.14)
***Schistosoma* spp. (number, %)**	
Urogenital schistosomiasis (301, 74.6%)	
S. haematobium (1)	263 (65%)
Schistosoma spp. (2)	38 (9.6%)
Hepatointestinal schistosomiasis (114, 28.3%)	
S. mansoni	79 (19.5%)
S. intercalatum/guineensis (3)	10 (2.5%)
Schistosoma spp. (4)	26 (6.4%)

SD: Standard deviation.

(1) 5 patients had co-infection with *S. mansoni* and 1 co-infection with *S. mansoni* and *S. intercalatum/guineensis*. Two patients also had positive rectal biopsies.

(2) Detection of *Schistosoma* eggs in biopsies from urogenital region: 35 in bladder biopsy (one of them also with *S. intercalatum/guineensis* in faeces), 2 in testicular biopsies, 1 in cervix biopsy.

(3) 1 patient had co-infection with *S. mansoni*, 1 with *S. haematobium* and *S. mansoni*, and 1 had a bladder biopsy showing *Schistosoma* eggs.

(4) Detection of *Schistosoma* eggs at hepatointestinal level: 24 in rectal biopsy (2 of them in addition with *S. haematobium*) and 2 in appendiceal biopsy.

In 263 patients (64.9%), *S. haematobium* eggs were demonstrated in urine, 79 (19.5%) had *S. mansoni* eggs in faeces and 10 (2.5%) had *S. intercalatum/guineensis* eggs in faeces.

Thirty-five patients were diagnosed after the detection of *Schistosoma* spp. eggs in bladder biopsies, 24 in rectal biopsies, 2 in testicular biopsies, 2 in appendicular biopsies and 1 in cervical biopsy.

Co-infection of different *Schistosoma* species was observed in 10 patients: 5 with *S. haematobium-S. mansoni* co-infection; 1 *S. intercalatum/guineensis-S. mansoni*; 1 *S. haematobium-S. mansoni-S. intercalatum/guineensis*; 1 patient had *Schistosoma* spp. eggs in bladder biopsy and *S. intercalatum/guineensis* in faeces; and 2 patients had *Schistosoma* spp. eggs in rectal biopsy and *S. haematobium* in urine.

Overall, serology was positive in 76% of cases (308/405) (Table 2). ELISA was used in 302 patients (74.6%), IHA in 9 (2.2%) and ICT in 94 (23.2%). Sensitivity values ranged from 44.4% for IHA to 94.7% for ICT/LDBIO. Serology was negative in 97 patients: 71 had *S. haematobium* infection, 11 had *S. mansoni* infection, one had triple *S. haematobium-S. mansoni-S. intercalatum/guineensis* co-infection and 14 were patients diagnosed by biopsy (10 bladder, 3 rectal and one cervical).

Table 2. Sensitivity of serological tests for the detection of *Schistosoma* infection.

Diagnostic Test	No. Positive/Total Number of Patients Tested (%)	No. Negative/Total Number of Patients Tested (%)
S. mansoni IgG-ELISA	215/302 (71.2)	87/302 (28.8)
Bilharziose Fumouze IHA®	4/9 (44.4)	5/9 (55.6)
Schistosoma ICT IgG-IgM®	89/94 (94.7)	5/94 (5.3)

To calculate the sensitivity of serological tests according to urogenital or hepatointestinal involvement, only patients with infection in one of the locations were considered. The results are shown in Table 3. For urogenital schistosomiasis, the sensitivity was 28.6% for IHA, 66.2% for ELISA/NovaLisaTM and 94.2% for ICT/LDBIO. For hepatointestinal schistosomiasis, sensitivities were 100% for IHA, 84.3% for ELISA/NovaLisaTM and 95% for ICT/LDBIO, respectively.

Table 3. Sensitivity of serological tests according to localisation.

Diagnostic Test	No. Positive/Total Number of Patients Tested (%)	
	Urogenital Schistosomiasis (N = 292)	Hepatointestinal Schistosomiasis (N = 104)
S. mansoni IgG-ELISA	143/216 (66.2)	70/83 (84.3)
Bilharziose Fumouze IHA®	2/7 (28.6)	1/1 (100)
Schistosoma ICT IgG-IgM®	65/69 (94.2)	19/20 (95)

To determine sensitivity according to *Schistosoma* species (Table 4), only patients with monoinfections were considered, also excluding patients with biopsy-diagnosed schistosomiasis. For *S. haematobium* the sensitivity was 28.6% with IHA, 65.7% with ELISA/NovaLisaTM and 94% with ICT/LDBIO. For *S. mansoni* the sensitivity was 100% for IHA, 82.1% for ELISA/NovaLisaTM and 93.3% for ICT/LDBIO. For *S. intercalatum/guineensis* the sensitivity was 100% for all techniques.

Table 4. Sensitivity of serological tests according to *Schistosoma* species.

Diagnostic Test	N° Positive/Total Number of Patients Tested (%)		
	S. haematobium	*S. mansoni*	*S. intercalatum/guineensis*
S. mansoni IgG-ELISA	119/181 (65.7)	46/56 (82.1)	3/3 (100)
Bilharziose Fumouze IHA®	2/7 (28.6)	1/1 (100)	-
Schistosoma ICT IgG-IgM®	63/67 (94)	14/15 (93.3)	4/4 (100)

4. Discussion

Based on the results of our study, we can affirm that ICT is the serological method analysed that shows the best sensitivity for the diagnosis of imported schistosomiasis in sub-Saharan migrants, both urogenital and hepatointestinal, in real clinical practice. As it is a rapid test, thus easy to perform and interpret, it could be recommended as a screening test in non-endemic regions.

Studies on the prevalence of schistosomiasis in migrants from endemic areas in Europe are scarce and show figures of seroprevalence much higher than those obtained by direct microscopy. Serre et al. reported a prevalence of schistosomiasis by microscopic examination of 9% in migrants living in shelters in Barcelona [26]. Salas-Coronas et al. found that in newly arrived African migrants in Spain, a direct diagnosis of schistosomiasis was made in 12.3% of the subjects while the seroprevalence was 32.2% [9]. In a similar study in Italy in refugees, 17.4% were diagnosed by microscopy and serology (ELISA) was positive in 27.6% of cases [27]. In a German study in unaccompanied minors, *Schistosoma* spp. eggs were visualized in 24.7% of cases [28]. In view of these data, we can state that despite the low sensitivity of microscopy, the prevalence of schistosomiasis in sub-Saharan migrants recently arrived in Europe is high.

The sensitivity of the different commercially available serological tests varies significantly depending on the technique used and the population under study. In relation to IHA, although the number of patients studied was small, we found a low sensitivity (44.4%) for the detection of *Schistosoma* spp. infection, which coincides with other studies such as that of Leblanc et al. that showed a sensitivity of 48% in a study conducted in children coming from an endemic area in the previous 12 months or in autochthones after a stay of at least 3 months in an endemic area and skin contact with fresh water during the journey [10]. Additionally, Yameny, in a cross-sectional study in Egypt designed to evaluate the efficacy of this technique compared to microscopy, obtained a 42% sensitivity using 50 *S. haematobium* positive samples and 50 negative ones [29]. However, Hinz et al., in a literature review, presenting performance data from a wide range of serological techniques, found sensitivity values of 73–94% for IHA Fumouze Diagnostics in imported schistosomiasis in travellers [21]. Kinkell et al., using frozen sera from 121 patients with various parasitic infections (with 37 cases of schistosomiasis among them) and 20 sera samples from healthy volunteers, obtained a sensitivity of 73% [22]. Van Gool et al., in a study evaluating various serological tests for the diagnosis of imported schistosomiasis in patients who had recently visited an African country endemic for schistosomiasis, reported a sensitivity for IHA of 86% and 94% depending on whether they considered the cut-off titre of 1:160 (suggested by the manufacturer) or 1:80, respectively [24].

The sensitivity obtained in our study for *S. mansoni* IgG-ELISA/NovaLisaTM was 71.2%. Beltrame et al. assessed the accuracy of several serological tests on the basis of microscopy results and obtained a sensitivity of 82% for ELISA [16]. In the study by Kinkell et al., the sensitivity was 64.9% for NovaTec ELISA [22].

The test that has shown the highest concordance with microscopy in the detection of schistosomiasis in our study was the Schistosoma ICT IgG-IgM immunochromatographic test, with a sensitivity of 94.7%. Schistosoma ICT IgG-IgM® is a rapid test that simultaneously detects IgG and IgM antibodies. In schistosomiasis, IgM levels peak at around 12–16 weeks after infection, while IgG peaks at around 20 weeks [18]. Therefore, the capacity of detecting both IgM and IgG could lead to a higher sensitivity of the test by detecting a higher proportion of recent infections and cases of acute schistosomiasis. Further studies in newly infected patients or in early stages of the disease would be necessary to explore how this affects the performance of the test. Several authors have reported results similar to ours. Beltrame et al. in a study in Italy with African migrants, using microscopy as the gold standard, reported a sensitivity of 94% [16]. Leblanc et al. found a sensitivity of 100% for Schistosoma ICT IgG-IgM® in their study [10]. In a recent publication by Hoermann et al., ICT showed a sensitivity of 100% in patients with confirmed schistosomiasis, irrespective of species as *S. mekongi* and *S. japonicum* infections were included in addition to *S. mansoni* and *S. haematobium* [25].

There have also been a few studies in schistosomiasis-endemic countries that have evaluated the diagnostic performance of Schistosoma ICT IgG-IgM. Two of them, one in Nigeria for the detection of urinary schistosomiasis using a Western blot (SCHISTO II WB IgG, LDBIO Diagnostics) as the gold standard [30] and one in Zambia for the detection of *S. mansoni* and *S. haematobium* infections using Kato-Katz and urine filtration [31], showed sensitivities of 94.9 % and 100 %, respectively.

When the results were analysed according to the different *Schistosoma* species, the sensitivity data obtained with IHA and ELISA were higher for *S. mansoni* infections (100% and 82.1%) than for *S. haematobium* (28.6% and 65.7%). However, no differences between species were found when ICT was used (93.3% for *S. mansoni* and 94% for *S. haematobium*). These data are similar to those found in Italy by Beltrame et al. [16], as they obtained sensitivity values of 84% for ELISA and 94% for ICT when considering only *S. mansoni*, and 79% for ELISA and 94% for ICT for *S. haematobium*. In our study, for IHA the data show a much better sensitivity in detecting *S. mansoni* than *S. haematobium*, although the results are limited by the small sample size. In any case, Van Gool et al. [24] suggested that reducing the cut-off titre from 1:160 to 1:80 would strongly increase the sensitivity of IHA (from 88% to 94.7% for *S. mansoni* and from 80% to 92% for *S. haematobium*) with only a slight drop in specificity.

Both *S. mansoni* IgG-ELISA/NovaLisaTM and Schistosoma ICT IgG-IgM showed 100% sensitivity in the diagnosis of *S. intercalatum*/*guineensis* infections, although the number of patients was very small in our series.

Due to the tropism of the different schistosome species, the sensitivity data for UG schistosomiasis are similar to those obtained for *S. haematobium* and those for HI schistosomiasis to those for *S. mansoni*.

Regarding the patients with a false negative serological result, 89.6% were obtained with ELISA, 5.2% with IHA and 5.2% with ICT. Consistent with the literature, the majority of cases, 72.3%, corresponded to *S. haematobium* infections [22]. Marchese et al. reported a proportion of false negatives of 17.5%, of which, 61.1% were *S. haematobium* infections [8]. This is probably due to the fact that most serological tests use antigens against *S. mansoni*. False negative results also occur more frequently in acute or recent infections when the presence of antibodies is not yet detectable (window period), in individuals with late seroconversion (up to 6 months delay) [22] or in those with a low level of antibody response [32]. For such reason, and in order to increase the sensibility of the diagnostic methods, some authors recommend the use of two or more assays in parallel [22]. In addition, in adults in endemic areas where the intensity of infection is generally lower than in young people [6], a decreasing antibody response may occur as a consequence of repeated exposures to schistosome cercariae [21,33]. On the other hand, using a confirmatory test in patients at risk of co-infection by other tissue-invasive helminths would allow to reduce the number of false-positives due to cross-reactivity.

The main limitations of our study derive, first, from its retrospective nature. Second, from the small sample size, especially in the case of IHA testing. Third, it has to be considered a potential technical variability in the procedure of the tests over the years analysed in our study. Fourth, it is possible that in some of the cases diagnosed by biopsy, eggs were already dead because of previous treatments received in home-countries or because of natural death. Nevertheless, in our case, invasive procedures in order to take biopsies (mainly cystoscopies and rectosigmoidoscopies) were merely indicated when active disease was clearly suspected. In the case of urogenital schistosomiasis, biopsies were taken only when hematuria or suggestive bladder nodules in the ultrasound examination were present. For intestinal schistosomiasis, only patients with abdominal pain, diarrhoea or rectal bleeding with no other alternative causes were considered for biopsy. Finally, another weakness is the unfortunate unavailability of archived biological samples to perform the three techniques simultaneously on all samples. On the other hand, the main strengths of our work are the establishment of sensitivity compared to the gold standard of microscopy, the fact that it is a study in real clinical practice, and the high number of patients with confirmed schistosomiasis in our series.

5. Conclusions

In conclusion, in view of the recommendations made by international organisations for schistosomiasis serological screening in non-endemic countries, it is necessary to establish protocols for detecting the disease with sensitive tools capable of diagnosing infected individuals in order to provide early treatment and prevent disease progression [16,18,34]. Our study shows that Schistosoma ICT IgG-IgM® immunochromatography has the best performance in real clinical practice, when compared to ELISA and IHA, in both *S. mansoni* and *S. haematobium* infections. Therefore, it could be the test chosen to screen at-risk individuals. Further studies involving a larger number of patients are needed to compare these serological techniques with others that increase the sensitivity of microscopy, such as molecular techniques or antigen detection tests.

Author Contributions: Conceptualization, M.P.L.-G. and J.S.-C.; methodology, M.P.L.-G. and J.S.-C.; formal analysis, M.P.L.-G. and J.S.-C.; investigation, M.P.L.-G., M.I.C.-B., A.B.L.-S., M.J.S.-P., N.C.-F., J.V.-V., J.B.-J. and J.S.-C.; data curation, M.P.L.-G.; writing—original draft preparation, M.P.L.-G. and J.S.-C.; writing—review and editing, M.P.L.-G., M.I.C.-B., A.B.L.-S., M.J.S.-P., N.C.-F., J.V.-V., J.B.-J. and J.S.-C.; funding acquisition, J.S.-C. All authors have read and agreed to the published version of the manuscript.

Funding: This study was conducted within the activities developed by the Red de Investigación Colaborativa en Enfermedades Tropicales–RICET (Project RD16/0027/0013 of the PN de I + D + I, ISCIII-Subdirección General de Redes y Centros de Investigación Cooperativa RETICS, cofinanced with FEDER funds—European Regional Development Fund—"A way to make Europe"/"Investing in your future"), Ministry of Health and Consumption, Madrid; by Proyecto de Investigación en Salud PI-0001–2019, Consejería de Salud de la Junta de Andalucía, Sevilla, cofinanced FEDER funds; and by the Research group PAIDI CTS582 of the Regional Ministry of Gender, Health and Social Policy of the Government of Andalusia.

Institutional Review Board Statement: The study was approved by the Research Ethics Committee of Almería (Spain) with the code 109/2022.

Informed Consent Statement: Not applicable.

Data Availability Statement: The data that support the findings of this study are available from the corresponding author upon reasonable request.

Conflicts of Interest: The authors declare no conflict of interest.

References

1. World Health Organization. Schistosomiasis. Available online: https://www.who.int/news-room/fact-sheets/detail/schistosomiasis (accessed on 19 December 2022).
2. Centers for Disease Control and Prevention. Schistosomiasis. Prevention & Control. Available online: https://www.cdc.gov/parasites/schistosomiasis/prevent.html (accessed on 19 December 2022).
3. European Centre for Disease Prevention and Control. Public Health Guidance on Screening and Vaccination for Infectious Diseases in Newly Arrived Migrants within the EU/EEA. Stockholm. 2018. Available online: https://www.ecdc.europa.eu/en/publications-data/public-health-guidance-screening-and-vaccination-infectious-diseases-newly (accessed on 19 December 2022).
4. Clerinx, J.; Soentjens, P. Schistosomiasis: Epidemiology and Clinical Manifestations. In *UpToDate*; Shefner, J.M., Ed.; UpToDate. Available online: https://www.uptodate.com/contents/schistosomiasis-epidemiology-and-clinical-manifestations (accessed on 1 October 2022).
5. Gray, D.J.; Ross, A.G.; Li, Y.S.; McManus, D.P. Diagnosis and management of schistosomiasis. *BMJ* **2011**, *342*, d2651. [CrossRef] [PubMed]
6. Colley, D.; Bustinduy, A.L.; Secor, W.E.; King, C.H. Human schistosomiasis. *Lancet* **2014**, *383*, 2253–2264. [CrossRef] [PubMed]
7. Comelli, A.; Riccardi, N.; Canetti, D.; Spinicci, M.; Cenderello, G.; Magro, P.; Nicolini, L.A.; Marchese, V.; Zammarchi, L.; Castelli, F.; et al. Delay in schistosomiasis diagnosis and treatment: A multicenter cohort study in Italy. *J. Travel Med.* **2020**, *27*, taz075. [CrossRef] [PubMed]

8. Marchese, V.; Beltrame, A.; Angheben, A.; Monteiro, G.B.; Giorli, G.; Perandin, F.; Buonfrate, D.; Bisoffi, Z. Schistosomiasis in immigrants, refugees and travellers in an Italian referral centre for tropical diseases. *Infect. Dis. Poverty* **2018**, *7*, 55. [CrossRef]

9. Salas-Coronas, J.; Cabezas-Fernández, M.T.; Lozano-Serrano, A.B.; Soriano-Pérez, M.J.; Vázquez-Villegas, J.; Cuenca-Gómez, J.Á. Newly Arrived African Migrants to Spain: Epidemiology and Burden of Disease. *Am. J. Trop. Med. Hyg.* **2018**, *98*, 319–325. [CrossRef]

10. Leblanc, C.; Brun, S.; Bouchaud, O.; Izri, A.; Ok, V.; Caseris, M.; Sorge, F.; Pham, L.L.; Paugam, A.; Paris, L.; et al. Imported schistosomiasis in Paris region of France: A multicenter study of prevalence and diagnostic methods. *Travel Med. Infect. Dis.* **2021**, *41*, 102041. [CrossRef]

11. Roure, S.; Valerio, L.; Pérez-Quílez, O.; Fernández-Rivas, G.; Martínez-Cuevas, O.; Alcántara-Román, A.; Viasus, D.; Pedro-Botet, M.L.; Sabrià, M.; Clotet, B. Epidemiological, clinical, diagnostic and economic features of an immigrant population of chronic schistosomiasis sufferers with long-term residence in a non-endemic country (North Metropolitan area of Barcelona, 2002–2016). *PLoS ONE* **2017**, *12*, e0185245. [CrossRef]

12. Asundi, A.; Beliavsky, A.; Liu, X.J.; Akaberi, A.; Schwarzer, G.; Bisoffi, Z.; Requena-Méndez, A.; Shrier, I.; Greenaway, C. Prevalence of strongyloidiasis and schistosomiasis among migrants: A systematic review and meta-analysis. *Lancet Glob. Health* **2019**, *7*, e236–e248. [CrossRef]

13. Mendoza-Palomar, N.; Sulleiro, E.; Perez-Garcia, I.; Espiau, M.; Soriano-Arandes, A.; Martín-Nalda, A.; Espasa, M.; Zarzuela, F.; Soler-Palacin, P. Schistosomiasis in children: Review of 51 imported cases in Spain. *J. Travel Med.* **2020**, *27*, taz099. [CrossRef]

14. Ramalli, L.; Mulero, S.; Noël, H.; Chiappini, J.D.; Vincent, J.; Barré-Cardi, H.; Malfait, P.; Normand, G.; Busato, F.; Gendrin, V.; et al. Persistence of schistosomal transmission linked to the Cavu river in southern Corsica since 2013. *Euro. Surveill.* **2018**, *23*, 18-00017. [CrossRef]

15. Salas-Coronas, J.; Bargues, M.D.; Lozano-Serrano, A.B.; Artigas, P.; Martínez-Ortí, A.; Mas-Coma, S.; Merino-Salas, S.; Vivas-Pérez, J.I.A. Evidence of autochthonous transmission of urinary schistosomiasis in Almeria (southeast Spain): An outbreak analysis. *Travel Med. Infect. Dis.* **2021**, *44*, 102165. [CrossRef]

16. Beltrame, A.; Guerriero, M.; Angheben, A.; Gobbi, F.; Requena-Mendez, A.; Zammarchi, L.; Formenti, F.; Perandin, F.; Buonfrate, D.; Bisoffi, Z. Accuracy of parasitological and immunological tests for the screening of human schistosomiasis in immigrants and refugees from African countries: An approach with Latent Class Analysis. *PLoS Negl. Trop. Dis.* **2017**, *11*, e0005593. [CrossRef]

17. Utzinger, J.; Becker, S.L.; van Lieshout, L.; van Dam, G.J.; Knopp, S. New diagnostic tools in schistosomiasis. *Clin. Microbiol. Infect.* **2015**, *21*, 529–542. [CrossRef] [PubMed]

18. Weerakoon, K.G.; Gobert, G.N.; Cai, P.; McManus, D.P. Advances in the Diagnosis of Human Schistosomiasis. *Clin. Microbiol. Rev.* **2015**, *28*, 939–967. [CrossRef]

19. McManus, D.P.; Dunne, D.W.; Sacko, M.; Utzinger, J.; Vennervald, B.J.; Zhou, X.N. Schistosomiasis. *Nat. Rev. Dis. Primers* **2018**, *4*, 13. [CrossRef] [PubMed]

20. Agbata, E.N.; Morton, R.L.; Bisoffi, Z.; Bottieau, E.; Greenaway, C.; Biggs, B.A.; Montero, N.; Tran, A.; Rowbotham, N.; Arevalo-Rodriguez, I.; et al. Effectiveness of Screening and Treatment Approaches for Schistosomiasis and Strongyloidiasis in Newly-Arrived Migrants from Endemic Countries in the EU/EEA: A Systematic Review. *Int. J. Environ. Res. Public Health* **2018**, *16*, 11. [CrossRef] [PubMed]

21. Hinz, R.; Schwarz, N.G.; Hahn, A.; Frickmann, H. Serological approaches for the diagnosis of schistosomiasis—A review. *Mol. Cell Probes* **2017**, *31*, 2–21. [CrossRef]

22. Kinkel, H.F.; Dittrich, S.; Bäumer, B.; Weitzel, T. Evaluation of eight serological tests for diagnosis of imported schistosomiasis. *Clin. Vaccine Immunol.* **2012**, *19*, 948–953. [CrossRef]

23. Song, H.B.; Kim, J.; Jin, Y.; Lee, J.S.; Jeoung, H.G.; Lee, Y.H.; Hong, S.T. Comparison of ELISA and Urine Microscopy for Diagnosis of *Schistosoma haematobium* Infection. *J. Korean Med. Sci.* **2018**, *33*, e238. [CrossRef]

24. Van Gool, T.; Vetter, H.; Vervoort, T.; Doenhoff, M.J.; Wetsteyn, J.; Overbosch, D. Serodiagnosis of imported schistosomiasis by a combination of a commercial indirect hemagglutination test with *Schistosoma mansoni* adult worm antigens and an enzyme-linked immunosorbent assay with *S. mansoni* egg antigens. *J. Clin. Microbiol.* **2002**, *40*, 3432–3437. [CrossRef]

25. Hoermann, J.; Kuenzli, E.; Schaefer, C.; Paris, D.H.; Bühler, S.; Odermatt, P.; Sayasone, S.; Neumayr, A.; Nickel, B. Performance of a rapid immuno-chromatographic test (Schistosoma ICT IgG-IgM) for detecting Schistosoma-specific antibodies in sera of endemic and non-endemic populations. *PLoS Negl. Trop. Dis.* **2022**, *16*, e0010463. [CrossRef] [PubMed]

26. Serre Delcor, N.; Maruri, B.T.; Arandes, A.S.; Guiu, I.C.; Essadik, H.O.; Soley, M.E.; Romero, I.M.; Ascaso, C. Infectious Diseases in Sub-Saharan Immigrants to Spain. *Am. J. Trop. Med. Hyg.* **2016**, *94*, 750–756. [CrossRef] [PubMed]

27. Beltrame, A.; Buonfrate, D.; Gobbi, F.; Angheben, A.; Marchese, V.; Monteiro, G.B.; Bisoffi, Z. The hidden epidemic of schistosomiasis in recent African immigrants and asylum seekers to Italy. *Eur. J. Epidemiol.* **2017**, *32*, 733–735. [CrossRef] [PubMed]

28. Theuring, S.; Friedrich-Jänicke, B.; Pörtner, K.; Trebesch, I.; Durst, A.; Dieckmann, S.; Steiner, F.; Harms, G.; Mockenhaupt, F.P. Screening for infectious diseases among unaccompanied minor refugees in Berlin, 2014–2015. *Eur. J. Epidemiol.* **2016**, *31*, 707–710. [CrossRef]

29. Yameny, A. The validity of indirect haemagglutination assay (IHA) in the detection of *Schistosoma haematobium* infection relative to microscopic examination. *J. Med. Life Sci.* **2019**, *1*, 57–64. [CrossRef]

30. Houmsou, R.S.; Wama, B.E.; Agere, H.; Uniga, J.A.; Jerry, T.J.; Azuaga, P.; Amuta, E.U.; Kela, S.L. Diagnostic accuracy of Schistosoma ICT IgG-IgM and comparison to other used techniques screening urinary schistosomiasis in Nigeria. *Adv. Lab. Med.* **2021**, *2*, 71–77.
31. Mudenda, J.; Hamooya, B.M.; Tembo, S.; Halwindi, H.; Siwila, J.; Phiri, M.M. Diagnostic accuracy of Schistosoma immunochromatographic IgG/IgM rapid test in the detection of schistosomiasis in Zambia. *J. Basic Appl. Zool.* **2022**, *83*, 4. [CrossRef]
32. Xie, S.Y.; Yuan, M.; Ji, M.J.; Hu, F.; Li, Z.J.; Liu, Y.M.; Zeng, X.J.; Chen, H.G.; Wu, H.W.; Lin, D.D. Immune responses result in misdiagnosis of *Schistosoma japonicum* by immunodiagnosis kits in egg-positive patients living in a low schistosomiasis transmission area of China. *Parasites Vectors* **2014**, *7*, 95. [CrossRef]
33. Colley, D.G.; Secor, W.E. Immunology of human schistosomiasis. *Parasite Immunol.* **2014**, *36*, 347–357. [CrossRef]
34. Sequeira-Aymar, E.; diLollo, X.; Osorio-Lopez, Y.; Gonçalves, A.Q.; Subirà, C.; Requena-Méndez, A. Recommendations for the screening for infectious diseases, mental health, and female genital mutilation in immigrant patients seen in Primary Care. *Aten. Primaria* **2020**, *52*, 193–205. [CrossRef]

Tropical Medicine and Infectious Disease

Article

Exploring Evolutionary Relationships within Neodermata Using Putative Orthologous Groups of Proteins, with Emphasis on Peptidases

Víctor Caña-Bozada [1], Mark W. Robinson [2], David I. Hernández-Mena [3] and Francisco N. Morales-Serna [4,*]

[1] Centro de Investigación en Alimentación y Desarrollo, Mazatlán 82112, Mexico
[2] School of Biological Sciences, Queen's University Belfast, 19 Chlorine Gardens, Belfast BT9 5DL, UK
[3] Centro de Investigación y de Estudios Avanzados, Instituto Politécnico Nacional, Unidad Mérida, Mérida 97310, Mexico
[4] Instituto de Ciencias del Mar y Limnología, Universidad Nacional Autónoma de México, Mazatlán 82040, Mexico
*   Correspondence: neptali@ola.icmyl.unam.mx

Abstract: The phylogenetic relationships within Neodermata were examined based on putative orthologous groups of proteins (OGPs) from 11 species of Monogenea, Trematoda, and Cestoda. The dataset included OGPs from BUSCO and OMA. Additionally, peptidases were identified and evaluated as phylogenetic markers. Phylogenies were inferred using the maximum likelihood method. A network analysis and a hierarchical grouping analysis of the principal components (HCPC) of orthologous groups of peptidases were performed. The phylogenetic analyses showed the monopisthocotylean monogeneans as the sister-group of cestodes, and the polyopisthocotylean monogeneans as the sister-group of trematodes. However, the sister-group relationship between Monopisthocotylea and Cestoda was not statistically well supported. The network analysis and HCPC also showed a cluster formed by polyopisthocotyleans and trematodes. The present study supports the non-monophyly of Monogenea. An analysis of mutation rates indicated that secreted peptidases and inhibitors, and those with multiple copies, are under positive selection pressure, which could explain the expansion of some families such as C01, C19, I02, and S01. Whilst not definitive, our study presents another point of view in the discussion of the evolution of Neodermata, and we hope that our data drive further discussion and debate on this intriguing topic.

Keywords: Platyhelminthes; monogenea; trematoda; cestoda; phylogenomic; parasites

Citation: Caña-Bozada, V.; Robinson, M.W.; Hernández-Mena, D.I.; Morales-Serna, F.N. Exploring Evolutionary Relationships within Neodermata Using Putative Orthologous Groups of Proteins, with Emphasis on Peptidases. *Trop. Med. Infect. Dis.* **2023**, *8*, 59. https://doi.org/10.3390/tropicalmed8010059

Academic Editor: Vyacheslav Yurchenko

Received: 13 December 2022
Revised: 10 January 2023
Accepted: 11 January 2023
Published: 12 January 2023

1. Introduction

The Neodermata are a group of Platyhelminthes that is made up of metazoan parasites of three classes: Monogenea (primarily ectoparasitic), Trematoda (endoparasitic flukes), and Cestoda (endoparasitic tapeworms). These parasites infect a variety of vertebrate hosts and can cause disease in humans, farmed fish, livestock, and domestic animals. Although the existence of the Neodermata as a monophyletic group is well established, the phylogenetic relationships between the neodermatan classes remain under debate [1]. The clarification of this is important for understanding the evolutionary origins of endo- and ectoparasitism as well as the complex life cycles of flatworms [1–4].

One of the major controversies is the monophyly of Monogenea [5,6]. This class is divided into two subclasses (Polyopisthocotylea and Monopisthocotylea), that form a well-supported monophyletic group [5,7]. However, some studies, based on morphological [8,9] and molecular characters [3,10], suggest that Polyopisthocotylea and Monopisthocotylea do not occupy the same clade.

One of the better supported hypotheses, based on molecular data, is the early divergence of Monogenea and the sister-group relationship of Trematoda and Cestoda [6].

Nonetheless, there is also some evidence that infers a closer relationship of Monogenea (especially Monopisthocotylea) with Cestoda [1,3,11]. Some of these discrepancies may be due to the variety and quality of sequence data used or because some taxa are poorly represented (e.g., Polyopisthocotylea). Moreover, platyhelminth mitochondrial genomes have a high substitution rate (around four times that of other bilaterian taxa) and so may not be ideal for this type of analysis [12,13]. Thus, analyses based on multiple nuclear genes may provide greater resolution of the phylogenetic relationships with Neodermata [12,14–17]. Currently, this task is facilitated by BUSCO and OMA, which are two leading programs for identifying putative orthologous groups of genes and proteins [18,19].

As an alternative to the use of complete genomes or transcriptomes, the use of a protein family with homology in a wide range of species could be considered in phylogenetic analyses [20]. In this context, given that neodermatan peptidases are important virulence factors that perform roles essential for parasitism, including feeding, migration, and avoidance/modulation of the host immune response [21–25], it would be interesting to explore their use as phylogenetic markers. Although characteristics such as lifestyle (free-living, ectoparasites, and endoparasites), host niche (gills, intestine, liver, etc.), and feeding (by degradation of host mucus, tissue, and blood or by direct uptake of digestion products) are variable among monogeneans, trematodes, and cestodes [3], secreted helminth peptidases (operating at the host-parasite interface) are under selective pressure from the host and thus may respond similarly and leave genuine phylogenetic signals across diverse lineages. For instance, positive selection can accelerate the fixation of advantageous mutations that enhance or refine the functions of the ancestral gene [26], as has been seen with the expansion of a family of cathepsins L with distinct but overlapping substrate specificities in the trematode, *Fasciola hepatica* [27]. Thus, the aim of this study was to explore the evolutionary relationships within Neodermata using putative orthologous groups of proteins (OGPs) retrieved from BUSCO and OMA, and to evaluate the use of peptidases as markers for phylogenetic studies.

2. Materials and Methods

2.1. Evolutionary Relationships of Neodermata

2.1.1. Phylogenetic Analysis of BUSCO and OMA OGPs

Two phylogenetic analyses were performed using putative OGPs obtained from the genomes (g), transcriptomes (t), or EST sequences of 11 species of Neodermata: the monopisthocotyleans *Rhabdosynochus viridisi* (t), *Scutogyrus longicornis* (t), *Gyrodactylus salaris* (g), and *Neobenedenia melleni* (EST) [28,29]; the polyopisthocotyleans *Protopolystoma xenopodis* (g) and *Eudiplozoon nipponicum* (t) [29,30]; the cestodes *Echinococcus multilocularis* (g), *Hymenolepis microstoma* (g), and *Taenia asiatica* (g) [29]; and the trematodes *F. hepatica* (g) and *Schistosoma mansoni* (g) [29]. The free-living platyhelminths *Bothrioplana semperi* (g), belonging to the order Bothrioplanida, and *Schmidtea mediterranea* (g), belonging to the order Tricladida, were used as outgroups [12,29]. The first analysis included single-copy OGPs retrieved from BUSCO v4 [31], using the core metazoan dataset, which contains 978 genes and the script BUSCO_phylogenomics.py with parameters "-supermatrix" and "-psc 70". The second analysis included simply OGPs (OMA Groups) obtained through OMA Standalone [32], using the script filter_groups.py [19]. OMA Groups contained a maximum of one representative gene per species. When multiple co-orthologs exist, OMA selected one to be in the OMA Group. Only the OGPs present in at least 10 species were included in the analyses.

The BUSCO proteins used in the phylogenetic analyses were annotated with the odb10 database [33], whereas the OMA proteins were annotated using BLASTp [34] against the UniProtKB/Swiss-Prot database (e-value $< 1 \times 10^{-4}$). In addition, the proteins were mapped to Gene Ontology (GO) terms using the PANNZER2 web server [35]. The visualisation and GO term enrichment analysis were performed in WEGO [36].

The OGPs were aligned with Muscle v3.8.31 [37], trimmed with trimAL [38] using the automated mode (-automated1), and concatenated. The best evolutionary models were obtained with the ModelFinder program [39]. Phylogenetic trees were constructed in IQ-TREE v1.6.12 [40], using the Shimodaira–Hasegawa-like approximate likelihood ratio test (SH-aLRT) (1000 replicates) and 1000 ultrafast bootstrap approximations to calculate the support values of the clades. Phylogenetic trees were also constructed in RAxML v8 [41], with 1000 bootstrap (Bs) iterations to calculate the support values of the clades. The trees were visualised with FigTree v1.4.2 (http://tree.bio.ed.ac.uk/software/figtree/, accessed on 1 November 2021). The trees were constructed using the models LG + F + R4 and LG + F + G + I for IQ-TREE and RAxML, respectively. Constrained trees were constructed in IQ-TREE (using the -g option), using alternative scenarios obtained from the literature as a guide (see Table 1) and the LG + F + R4 model. To determine whether the topology found in this study was significantly better than constrained trees, tree topology tests were performed in IQ-TREE using the LG + F + R4 evolutionary model with the options: -au, -zb.

Table 1. Tree topological test for four evolutionary scenarios of neodermatans, based on OGPs of two datasets (BUSCO and OMA).

Scenario	Constrained Tree	logL	deltaL	bp-RELL	*p*-KH	*p*-SH	c-ELW	*p*-AU	Reference
				BUSCO OGPs					
1	((Monopisthocotylea, Cestoda), (Polyopisthocotylea, Trematoda))	−504,202.9113	0	0.941+	0.943+	1+	0.938+	0.947+	Present study
2	((Monopisthocotylea, Polyopisthocotylea), (Cestoda, Trematoda))	−504,265.679	62.768	0.0137−	0.0567+	0.0567+	0.0205−	0.0542+	[6]
3	(((Monopisthocotylea, Polyopisthocotylea), Cestoda), Trematoda)	−504,265.679	62.768	0.0208−	0.0567+	0.0567+	0.0205−	0.0538+	[42]
4	(((Trematoda, Cestoda), Polyopisthocotylea), Monopisthocotylea)	−504,265.679	62.768	0.0248+	0.0567+	0.0567+	0.0205+	0.0534+	[3]
				OMA OGPs					
1	((Monopisthocotylea, Cestoda), (Polyopisthocotylea, Trematoda))	−2,015,132.409	0	0.987+	0.987+	1+	0.986+	0.986+	Present study
2	((Monopisthocotylea, Polyopisthocotylea), (Cestoda, Trematoda))	−2,015,328.208	195.8	0.0029−	0.0131−	0.0131−	0.00454−	0.0142−	[6]
3	(((Monopisthocotylea, Polyopisthocotylea), Cestoda), Trematoda)	−2,015,328.208	195.8	0.0044−	0.0131−	0.0131−	0.00454−	0.0142−	[42]
4	(((Trematoda, Cestoda), Polyopisthocotylea), Monopisthocotylea)	−2,015,328.208	195.8	0.0062−	0.0131−	0.0131−	0.00454−	0.0144−	[3]

deltaL: logL difference from the maximal logl in the set. bp-RELL: bootstrap proportion using RELL method [43]. *p*-KH: *p*-value of one-sided Kishino–Hasegawa test [44]. *p*-SH: *p*-value of Shimodaira–Hasegawa test [45]. c-ELW: Expected Likelihood Weight [46]. *p*-AU: *p*-value of approximately unbiased (AU) test [47]. Plus signs denote the 95% confidence sets. Minus signs denote significant exclusion. All tests performed 10,000 resamplings using the RELL method.

2.1.2. Phylogenetic Analysis of Peptidases

Peptidase and peptidase inhibitors were identified in the predicted proteins of the aforementioned 11 species of Neodermata and two free-living platyhelminth species. To this end, we followed the recommendations of Rawlings et al. [48]. The predicted proteins were aligned against the MEROPS v12.3 database [48] using BLASTp (e-value $< 1 \times 10^{-4}$). To avoid the overestimation of sequences, only the protein isoform with the best e-value was considered in the analysis. It was not possible to identify the isoforms in all the files, so CD-HIT v4.8.1 [49] with 90% sequence identity was run for all the species, following Ji et al. [50]. The retained proteins were aligned against the NCBI non-redundant protein database for detecting potential contaminant sequences from bacteria, viruses, fish, or tetrapods. The resulting sequences were classified according to the MEROPS database and included aspartic, cysteine, glutamic, metallo, asparagine, mixed, serine, threonine, unknown peptidases, and inhibitor classes. Classically secreted proteins were identified with SignalP v4.1 [51], using default settings for eukaryotes. In addition, SecretomeP [52] with default options for mammalian organisms and filtered by NN-scores larger than 0.9 was used to identify non-classically secreted proteins.

The peptidases and peptidase inhibitors were submitted to OMA Standalone to obtain orthologous groups (OGs) and to make comparisons between neodermatans. In order to improve the retrieval of these OGs, the phylogenetic relationships data (in Newick format) were also submitted to OMA.

A multilocus phylogenetic analysis of peptidases was performed using the OMA Groups. This analysis included the peptidases present in at least 10 of the 13 species of platyhelminths. Each group of peptidases was aligned with MAFFT, using the parameters "-maxiterate 1000 and -localpair", trimmed with the parameter "-gappyout", and posteriorly concatenated. The selection of the best evolutionary models, the construction, and the visualisation of the trees were performed as described above. The trees were constructed using the models LG + F + R4 and LG + F + G + I for IQ-TREE and RAxML, respectively.

2.1.3. Network Analysis of Peptidase OGs

A network analysis was performed using the OGs matrix of Hierarchical Orthologous Groups obtained from OMA Standalone. The Gephi v0.9.2 software [53] was used to construct a directed network. The communities (internal subdivisions) of the network were obtained following the modularity optimization method, based on the number of peptidases in each OG, which can be applied to weighted and directed graphs [54]. Some resolution values were evaluated, and the least variable resolution was chosen. The final parameters used in the analysis were resolution of 1.0, edge weights, and randomization [54]. In a network, the communities (clusters or modules) can be defined as groups of vertices that have common properties and/or perform similar roles, and that occur when the number of edges inside the subdivisions is higher than the expected number of internal edges that the same subgraph would have in the null model [55]. The values of modularity can fall between −1 and 1 and measure the density of links inside communities as compared to links between communities [54]. Higher positive values are linked to high densities of links inside communities and lower links between communities (better clustering), negative values indicate the opposite, and a zero value expresses link randoms in the cluster of networks [56]. The OGs associated with the formation of each cluster were retrieved.

2.1.4. Hierarchical Grouping Analysis in Principal Components of OGs

This analysis was performed with the R package "FactoMineR" [57] using the OGs matrix of Hierarchical Orthologous Groups. To prevent variables with large values from becoming dominant, the data were standardised with the option "scale.unit = TRUE", which scaled the values to unit variance. Thereafter, the data were subjected to principal component analysis (PCA) to reduce the dimension of the data (principal components). Finally, the hierarchical clustering, using Euclidean distance matrices and Ward's hierarchical clustering algorithm, was performed to obtain the clusters. Ward's hierarchical clustering algorithm uses an ANOVA approach for calculating distances between clusters [58]. The hierarchical clustering analysis was performed using the parameter "nb.clust = 4" to obtain the same number of clusters observed in the phylogenomic analysis. The genes associated with each cluster were retrieved (*p*-value < 0.05) and compared with those obtained from the network analysis to obtain the main subfamilies of peptidases influencing the formation of the cluster. The information about the main subfamilies was visualised in a circus plot, using the R package "circlize" [59].

2.2. Mutation Rates in Secreted and Non-Secreted Peptidases

The paired OGs obtained from OMA Standalone (PairwiseOrthologs output in OMA) were analysed to estimate the rate of synonymous (Ks) and non-synonymous (Ka) substitutions among species of each class of platyhelminths. Briefly, each OG among species of the same class was aligned using MAFFT v7.31 [60] (-auto). The amino acid alignment was used as a guide to generate the nucleotide sequence alignment using ParaAT [61]. Then, the alignments were used to calculate Ks and Ka for each OG with the software Codeml (CodonFreq = 2, runmodel = −2) of the PAML package v4.8 [62]. The Ks and Ka were obtained using the pipeline PANAS [63]. Because the peptidases present high variability of nucleotides, all Ks values were retained, in contradiction with other analyses [64]. The Kruskal–Wallis test was used to evaluate significant differences (p-value < 0.05) of Ks and Ka between (1) single-copy OGs vs. multiple-copy OGs; (2) single-copy OGs secreted proteins vs. single-copy OGs non-secreted proteins; and (3) multiple-copy OGs secreted proteins and multiple-copy OGs non-secreted proteins.

2.3. Classification of the C01A and S01C Peptidase Subfamilies

The sequences of the C01A and S01C peptidase subfamilies were submitted to MOTIF Search (https://www.genome.jp/tools/motif/MOTIF.html, accessed on 1 December 2021) to retrieve the region of the Pfam domain Peptidase_C1 (PF00112) for the subfamily C01A, and Trypsin (PF00089) for the subfamily S01C. The members of these subfamilies were classified through phylogenetic analyses. Trees were constructed as described above, using only the approximate likelihood ratio test (1000 replicates), and visualised and annotated with FigTree and the ITOL web server v5 [65]. Bootstrap values above 80% were considered as significant. The phylogenetic analysis was performed using reference proteins of platyhelminths from the MEROPS database. In addition, the alignment of the sequences was used to identify the S2 subsite residues belonging to cathepsins L (positions 67, 157, and 205; papain numbering) in the subfamily C01A [27,66].

3. Results

3.1. Evolutionary Relationships of Neodermata

3.1.1. Phylogenetic Analyses of BUSCO and OMA OGPs

The phylogenetic analyses were performed using 137 BUSCO OGPs (Supplementary File S1 and Table S1) and 479 OMA OGPs (Supplementary File S2 and Table S2). In the phylogenetic analysis using BUSCO OGPs, Monopisthocotylea appeared as the sister-group of Cestoda (Bs = 62%), whereas Polyopisthocotylea appeared as the sister-group of Trematoda (Bs = 64%). In turn, each of the classes and subclasses formed well-supported monophyletic groups (Bs = 100%) (Figure 1A). The tree constructed using OMA OGPs showed the same topology, with high level support in all the nodes (Bs > 97%, Figure 1B). The tree topology test revealed that by using BUSCO OGPs no scenario was rejected, whereas by using OMA OGPs only scenario 1 was not rejected (p-AU > 0.05) (Table 1). Constrained trees are shown in Supplementary Figure S1. The high Bs values observed in the tree based on OMA OGPs are possibly due to the use of a large dataset, because the same analysis with less data (200 OGPs) grouped Monopisthocotylea and Cestoda with support of 59%, and Polyopisthocotylea and Trematoda with 84% (tree not shown).

Details of the annotation of BUSCO and OMA OGPs is shown in Supplementary Tables S1 and S2.

3.1.2. Phylogenetic Analysis of Peptidases

A total of 3960 peptidases were identified in the 13 species of platyhelminths, distributed in 80 families across the five major classes (Supplementary Tables S3 and S4, Figures S2A and S3 and File S3). Serine peptidases were the most abundant class in monopisthocotyleans, whereas cysteine peptidases were the most abundant in polyopisthocotyleans, cestodes, trematodes, and free-living platyhelminths, except for *S. mansoni* and *T. asiatica*, where metallo peptidases were the most abundant. The most abundant peptidase families in all the platyhelminths studied were C19 (ubiquitinyl hydrolases), S01 (chymotrypsin family), S09 (prolyl oligopeptidases), C01 (papain family), and T01 (proteasome family) (Figure 2 and Supplementary Table S5). A total of 667 peptidase inhibitors belonging to

17 families were identified in all the platyhelminths (Supplementary Tables S3 and S5 and Figure S2B), with the family I02 (Kunitz) being the most abundant. Details of the identified peptidases are presented in Supplementary Results.

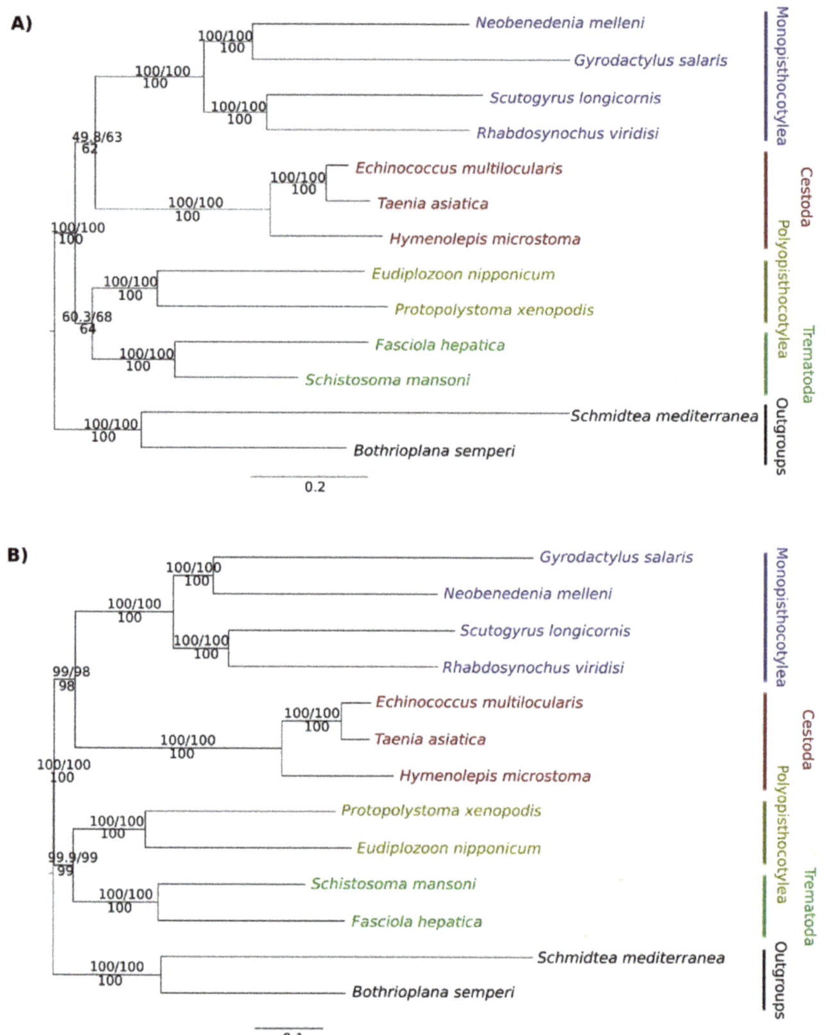

Figure 1. Phylogenetic trees of 13 platyhelminth species, based on (**A**) single-copy BUSCO OGPs and (**B**) simply OMA OGPs. Values above represent SH-aLRT support (%)/ultrafast bootstrap support (%) and below represent Maximum Likelihood rapid bootstrap support. The free-living platyhelminths *Bothrioplana semperi*, belonging to the order Bothrioplanida, and *Schmidtea mediterranea*, belonging to the order Tricladida, were used as outgroups.

SignalP predicted 565 peptidases (14.27%) and 237 peptidase inhibitors (35.53%) as classically secreted proteins (Figure 3A and Supplementary Table S6). The I02, I93, C01, M12, and S01 families presented a high proportion of proteins with signal peptides in comparison with other families (>40% of proteins with predicted N-terminal signal peptides) (Figure 3B). In addition, 79 peptidases (2%) and 18 peptidase inhibitors (2.7%) were predicted as non-classically secreted proteins, i.e., proteins lacking a signal peptide for secretion via the ER/Golgi pathway that exit the cell via atypical means [52].

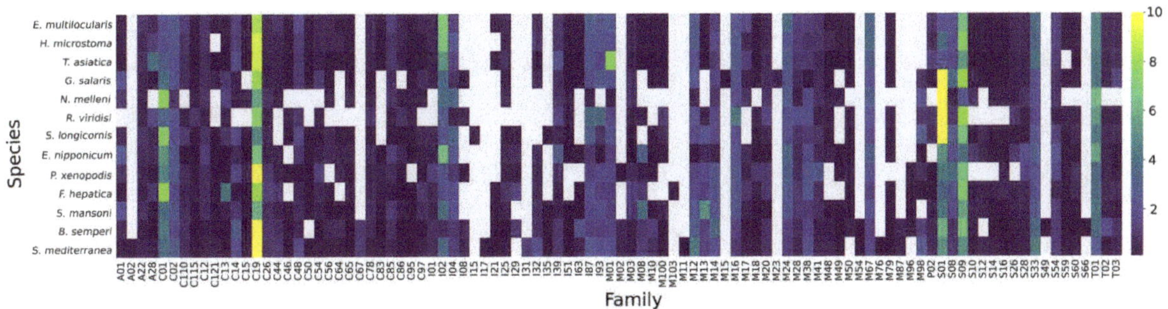

Figure 2. Heat map showing the proportion of peptidase and inhibitor families in each platyhelminth species.

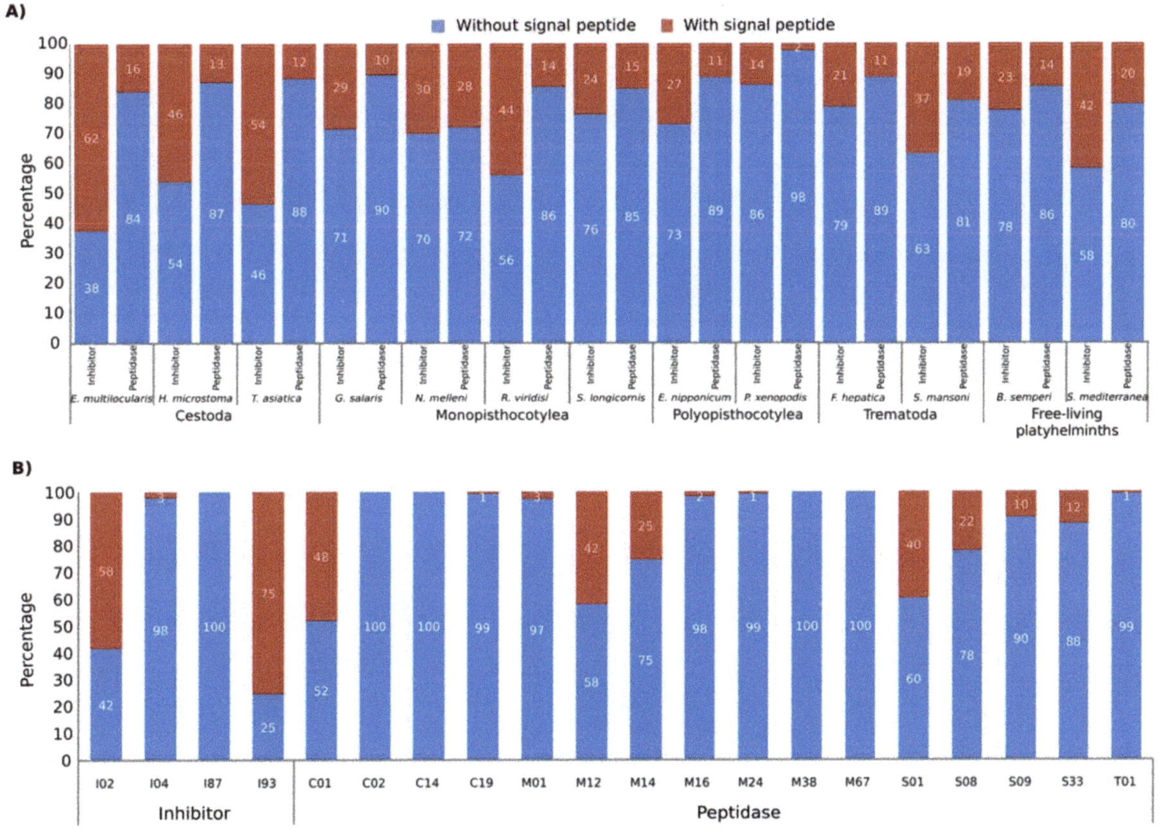

Figure 3. Peptidases and inhibitors with predicted signal peptides in the platyhelminth species analysed. (**A**) The proportion of peptidases and inhibitors with/without predicted signal peptides for secretion. (**B**) Proportion of secreted protein in the top 20 most abundant peptidase and inhibitor families.

A total of 531 OGs (HOGFasta output in OMA) of peptidases and peptidase inhibitors were found in the platyhelminths analysed (Supplementary Table S7 and File S4); however, only 85 OGs were present in at least 10 species (Supplementary File S5). The phylogenetic analysis was performed based on these 85 OGs. The multilocus phylogenetic tree of peptidases showed the same topology as the phylogenetic trees of the BUSCO and OMA OGPs (Figures 1 and 4A). The monopisthocotyleans were clustered with high support (100%) and had a sister-group relationship with cestodes, although with low support

(Bs = 48%). The polyopisthocotyleans formed a well-supported group (100%) and a well-supported sister-group relationship with trematodes (Bs = 97%). Cestodes and trematodes also formed their respective well-supported monophyletic groups (Bs = 100%). Details of the OGs annotation are shown in Supplementary Tables S8 and S9.

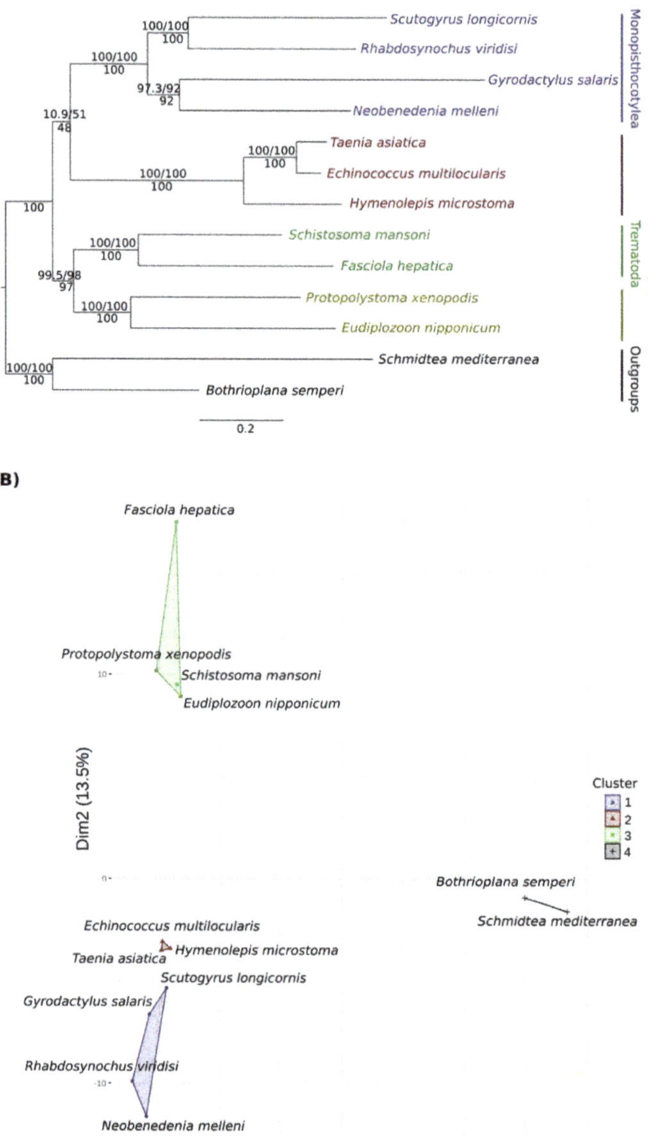

Figure 4. Peptidases and inhibitors of 13 platyhelminth species analysed. (**A**) Phylogenetic tree with values above representing SH-aLRT support (%)/ultrafast bootstrap support (%) and below representing Maximum Likelihood rapid bootstrap support. The free-living platyhelminths *Bothrioplana semperi*, belonging to the order Bothrioplanida, and *Schmidtea mediterranea*, belonging to the order Tricladida, were used as outgroups. (**B**) Formation of clusters according to their peptidases and inhibitors OGs using hierarchical grouping analysis in principal components.

3.1.3. Network Analysis and Hierarchical Grouping Analysis in Principal Components of Orthologous Groups

To identify the OGs most related to each class of platyhelminths, network analysis and a hierarchical grouping analysis in principal components (HCPC) was performed using the peptidases and peptidase inhibitors. The network analysis showed four clusters; a first cluster was formed by monopisthocotyleans, a second by cestodes, a third by both trematodes and polyopisthocotyleans, and a fourth by free-living species (Supplementary Figure S4). The network analysis clusters 134 OGs with monopisthocotyleans, 96 with cestodes, 176 trematodes and polyopisthocotyleans, and 125 OGs with free-living platyhelminths (Supplementary Table S10).

The HCPC analysis showed a similar clustering as that observed in the network analysis (Figure 4B). The cluster of monopisthocotyleans was associated with 27 OGs, cestodes with 54 OGs, trematodes + polyopisthocotyleans with 72 OGs, and free-living species with 114 OGs. These associations were significant and positive (Supplementary Table S11). OGs associated with each group in both the network and HCPC analysis are shown in Supplementary Table S12.

The main families that contributed to the formation of the four clusters in the network and HCPC analyses were C19, S09, S01, C02, C01, S01, I02, and I87 (Figure 5 and Supplementary Table S12). Particularly, the family I02 was important for the grouping of cyclophyllidean cestodes, the subfamily C01A for trematodes + polyopisthocotyleans, the subfamily S01C for monopisthocotyleans, and the subfamilies S09X and C02A for the free-living species used. Given their biological importance for neodermatan parasites [23,67,68], the C01A (cathepsin) and S01C (cercarial elastase) peptidases were further classified (see below).

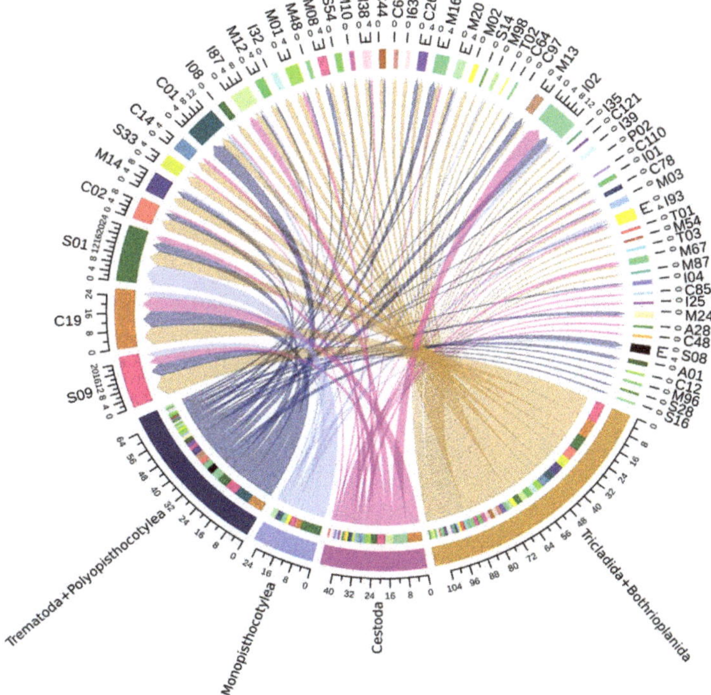

Figure 5. Circus chart representing the OGs of peptidase and inhibitor families that most contribute to the formation of groups in the HCPC and network analyses.

3.2. Mutation Rate in Secreted and Non-Secreted Peptidases

3.2.1. Single-Copy Orthologous Groups vs. Multiple-Copy Orthologous Groups

While Ka was higher in the multiple-copy OGs than in single-copy OGs (Supplementary Figure S5A), Ks was similar between OGs of single-copy and multiple-copy OGs (Supplementary Figure S5B). The lowest Ka and Ks in both single-copy and multiple-copy OGs were observed between cestode species (Supplementary Figure S6), although this may be because the three species of cestodes belong to the same order. The highest Ka in multiple-copy OGs was observed between monopisthocotylean species (Supplementary Figures S5A and S6), while the highest Ks in single-copy OGs were similar between platyhelminths, except in cestodes. All values of Ka and Ks are presented in Supplementary Table S13.

3.2.2. Single-Copy Secreted Protein Orthologous Groups vs. Single-Copy Non-Secreted Proteins Orthologous Groups

The analyses of secreted and non-secreted proteins were performed in both single-copy and multiple-copy OGs. Ka was higher in secreted proteins than in non-secreted proteins in most paired comparisons (Figure 6A). Ks in single-copy OGs was similar between secreted proteins and non-secreted proteins (Figure 6B) and between parasitic and free-living species.

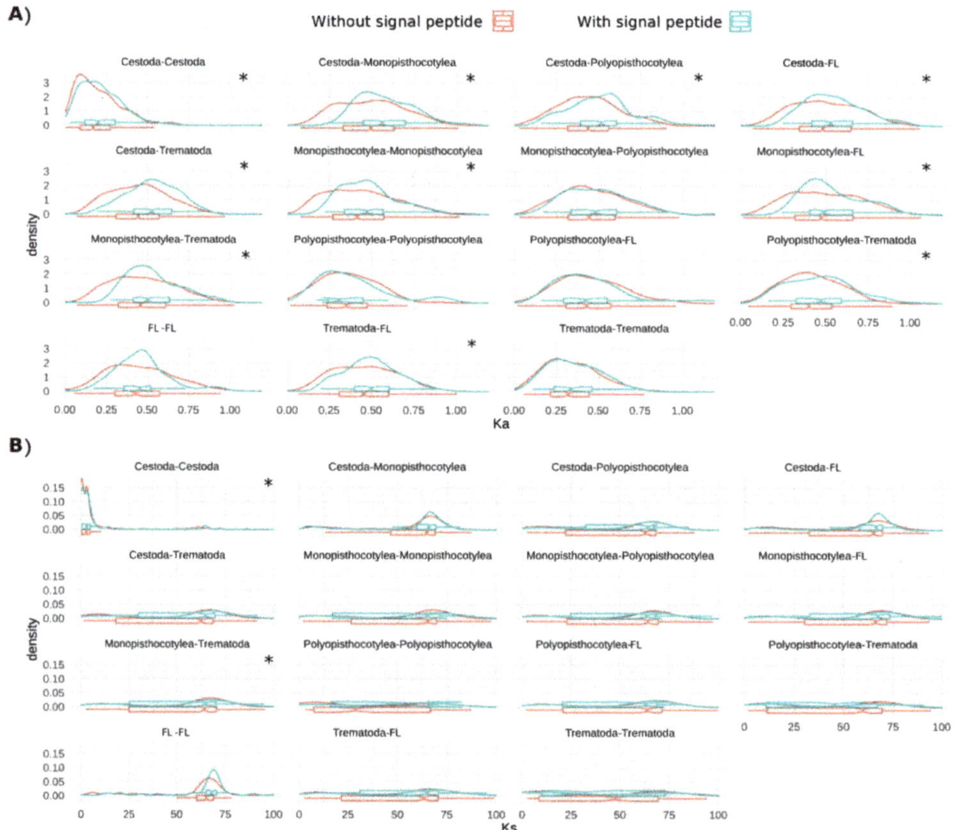

Figure 6. The distribution of (**A**) Ka and (**B**) Ks in single-copy OGs of peptidases and inhibitors of platyhelminths with/without predicted signal peptides. FL indicate the free-living platyhelminths *Bothrioplana semperi* and *Schmidtea mediterranea*. * indicates $p < 0.05$.

3.2.3. Multiple-Copy Orthologous Groups Secreted Proteins vs. Multiple-Copy Orthologous Groups Non-Secreted Proteins

Ka and Ks in multiple-copy OGs were higher in secreted proteins than in non-secreted proteins in most paired comparisons (p-value < 0.05) (Figure 7). The comparison between trematode species was the only one that showed a Ka higher in non-secreted proteins than in secreted proteins (Figure 7A).

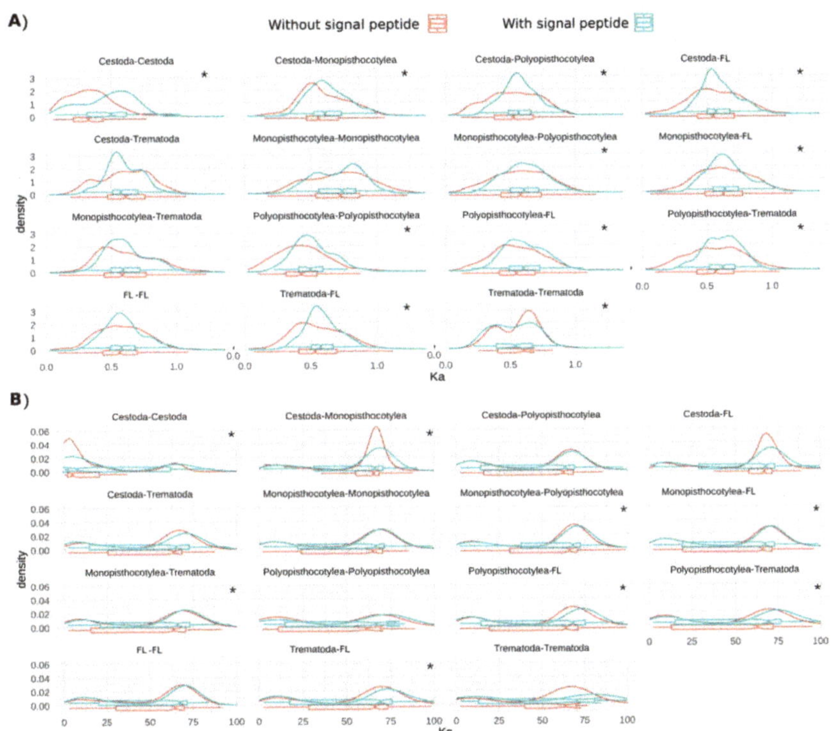

Figure 7. The distribution of (**A**) Ka and (**B**) Ks in multiple-copy OGs of peptidases and inhibitors of platyhelminths with/without predicted signal peptides. FL indicate the free-living platyhelminths *Bothrioplana semperi* and *Schmidtea mediterranea*. * indicate $p < 0.05$.

3.3. Classification of the C01A Peptidase Subfamily and S2 Active Subsite Residues

Members of the C01A papain-like peptidases were classified into two clades and three subclades as follows: subclade 1.1, formed by putative Cathepsin L proteins; subclade 2.1, formed by putative Cathepsin B proteins; and subclade 2.2, formed by putative dipeptidyl-peptidase proteins (Cathepsin C) (Table 2, Supplementary Table S14 and Figure S7). One subgroup within the cathepsin L (subclade 1.1) was formed only by peptidases of neodermatans.

The specific arrangement of amino acids that create the S2 subsite within the active site of cathepsin peptidases largely determines the specificity of the enzymes [69,70]. Studies using functionally active recombinant molecules showed that changes at residue positions 67, 157, and 205 (papain numbering) had the most significant impact on substrate specificity in *F. hepatica* cathepsins L [66,71,72]. Thus, the amino acid composition of the S2 subsite, at these three positions, was compared among the various platyhelminth cathepsin L sequences (Supplementary Figure S8 and Table S15). Leucine was the most represented amino acid at positions 67 and 157 in all species examined, while Methionine occurred most frequently at position 205 in all species, except *E. nipponicum*, *F. hepatica*, and *N. melleni*.

Table 2. Main members of the C01A and S01C subfamilies are classified in the phylogenetic analysis.

Species	C01A Subfamily			S01C Subfamily
	Cathepsin B	Cathepsin C	Cathepsin L	Cercarial Elastase
E. multilocularis	2	0	4	0
H. microstoma	2	0	4	0
T. asiatica	3	0	7	0
G. salaris	3	3	7	19
N. melleni	10	1	5	37
R. viridisi	2	1	3	8
S. longicornis	6	2	14	0
E. nipponicum	2	5	12	1
P. xenopodis	6	3	7	0
F. hepatica	17	0	16	1
S. mansoni	6	1	6	5
B. semperi	9	1	16	0
S. mediterranea	2	2	20	0
Total	70	19	121	71

We identified nine cathepsin L belonging to cestodes, trematodes, and monopisthocotyleans, which contained a tyrosine at position 67 (TASs01205g12006m00001, EmuJ_000989200.1, TASs00007g01862m00001, HmN_000323300, Smp_187140.1, TASs00112g07689m00001, maker-scf7180006948404-augustus-gene-0.32-mRNA-1, TRINITY_DN2691_c0_g1_i1__g.61367, and TRINITY_DN5043_c0_g1_i3__g.56686). Of these, the proteins belonging to cestodes and trematodes showed the same amino acids at positions 67 and 157 (tyrosine and leucine, respectively) that occur in the FhCL2 cathepsin (Supplementary Table S15). Additionally, we identified three cathepsin L sequences belonging to monogeneans (E_nip_trans_37948_m. 257852, TRINITY_DN10437_c1_g1_i2__g.43598, and TRINITY_DN3178_c0_g1_i10.p1) with the same amino acid distribution in the S2 subsite as FhCL5 (leucine at all three positions).

3.4. Classification of the S01C Peptidase Subfamily

The subfamily S01C was classified into two clades (clades 1 and 2) (Table 2, Supplementary Figure S9 and Table S14). The peptidases of clade 1 were formed by proteins of pathogenic neodermatans and clustered with the cercarial elastase peptidase (Smp_330280.1 = MER0003620) of *S. mansoni*. This group was overrepresented by monopisthocotylean proteins and lacked members from cestodes and free-living species. The clade 2 was comprised of proteins from all platyhelminth classes.

4. Discussion

In the present study, phylogenetic analyses based on OGPs from BUSCO and OMA, and peptidase OGs were performed to investigate the evolutionary relationships within Neodermata. Our dataset included 11 species of Neodermata, with members of both subclasses of Monogenea (Polyopisthocotylea and Monopisthocotylea), which were generally underrepresented in previous studies. The most important results at the phylogenetic level were the following: (1) It is corroborated again that Monogenea is not a monophyletic group because its subclasses are nested in different clades; (2) Monopisthocotylea and Cestoda were grouped in the same clade; and (3) Polyopisthocotylea and Trematoda were also grouped in the same clade. These findings are discussed below.

4.1. Non-Monophyly of Monogenea

The non-monophyly of the monogeneans has been the subject of discussion for many years, and this has intensified since molecular data began to be used to infer its phylogenetic position. In principle, the use of different sources of information (e.g., morphological vs. molecular), of different molecular markers (e.g., nuclear vs. mitochondrial and DNA vs. amino acid sequences), of different loci (e.g., a single gene vs. multigenes), of total evidence (i.e., morphology + molecular), and of different taxa included in the analyses (e.g., exclusion of some groups of monogeneans) has led to the formulation of inconsistent phylogenetic hypotheses (Supplementary Table S16).

The pioneering studies of Lambert [73], Brooks et al. [74], Ehlers [75], Rohde [76], Justine [8], and Boeger and Kritsky [7], offered a first approximation of the position and phylogenetic relationships of Monogenea, using morphological characters. However, it was Justine [77] who questioned the monophyly of Monogenea, based on spermatological analysis which suggested there was no synapomorphy that supported the monophyly of the group. Whilst surprising, this finding was also corroborated by the first phylogenetic analyses to use DNA sequences [78–81], although a subsequent molecular study obtained the opposite result, that Monogenea is monophyletic [42]. A later study combining molecular and morphological characters again suggested the monophyly of Monogenea [2], and another phylogenetic study with only morphological characters suggested that Monopisthocotylea and Polyopisthocotylea share the following synapomorphies: (1) larva with three zones ciliated; (2) two pairs of pigmented eyes in larvae and adults, with retention of the number and distribution of larval eyes in the adult; (3) a pair of ventral anchors; and (4) an egg filament [5,6]. Despite this, the monophyly of Monogenea continues to be questioned based on the increasing incorporation of new molecular data, such as more and new loci (ribosomal and mitochondrial genes), complete mitochondrial genomes (mitogenomes), genomes, and transcriptomes [3,4,10,82]. An alternative hypothesis that tries to explain why Monogenea cannot be consistently resolved as a monophyletic group is that once the Polyopisthocotylea separated from Monopisthocotylea, they had a rapid molecular divergence, accumulating a large number of mutations, which ended in homoplasies that can generate noise in phylogenetic inference [83]. However, studies using large DNA and amino acid sequence datasets from different genomic regions of the parasites, such as the one presented in this study, are convincing in obtaining independent and well-supported clades of the monogeneans analysed. Although the most recent studies (including ours) do not have a wide representation of species of both subclasses of monogeneans, it is clear that even with few taxa, large sets of molecular data indicate that Monopisthocotylea and Polyopisthocotylea do not nest in the same clade, which is consistent with our findings using OGPs. Therefore, all the studies that agree with the non-monophyly of Monogenea support the idea that the Monopisthocotylea and the Polyopisthocotylea do not share a recent common ancestor (despite similarities in their life cycle and in some morphological characteristics), and that they diverged from different ancestors that were more closely related to the other groups of Neodermata.

4.2. Monopisthocotylea + Cestoda

In the present study, the monophyly of Monopisthocotylea plus Cestoda was only well supported with OMA OGPs; nonetheless, the high support was not observed when the analysis was repeated using less data. In addition, our tree topology test based on BUSCO OGPs showed that a scenario in which Cestoda and Trematoda are grouped in a clade cannot be rejected. Other studies based on molecular data have already suggested a sister relationship between monogeneans and cestodes [10,12], which was also indicated by the presence of cercomers, or hooks, on the posterior end of larvae [84]. However, Lockyer et al. argued that the grouping of Monogenea with Cestoda by the cercomers is unreliable [6], and phylogenetic analyses using rRNA genes or mitochondrial genomes did not support this sister-group relationship [3,6].

Additionally, another morphological characteristic shared between monopisthocotyleans and cestodes (at least with *Gyrocotylidea*) is the presence of anterior nephridiopores [12,85]. It is interesting that our data provide some evidence in favour of a sister-group relationship between Cestoda and Monopisthocotylea when using peptidases. Because rRNA genes or mitochondrial genomes (used in conflicting studies) are sensitive to sequence alignment methods and are subject to rapid substitution rates [12], it is possible that peptidases may have value as phylogenetic markers in future studies.

4.3. Polyopisthocotylea + Trematoda

The monophyly of Polyopisthocotylea and Trematoda found in this study was also detected in the phylogenetic analysis of 202 single-copy OGPs of a number of helminth species (although this study did not include monopisthocotylean species) [82] and in a previous study using 28S DNA sequences [81]. It should also be noted that in the latter study, the clade Monopisthocotylea + Cestoda was also obtained. Our tree topology test based on BUSCO OGPs indicated that different scenarios are possible; nonetheless, the monophyly of Polyopisthocotylea and Trematoda was well supported by peptidase OGs and OMA OGPs, even when a relatively smaller dataset was used.

Trematoda have a variety of diets, including host blood and epithelia, with digestion in the gut being largely extracellular, while the Polyopisthocotylea feed on blood and have intracellular digestion [83]. Previously it had been suggested that the ancestor of the Trematodes could have been a Polyopisthocotylea-like sanguinivore, because a polyopistho-cotylean was the sister species of the clade formed by Trematoda + Cestoda [3]. However, our findings do not support this hypothesis, because Polyopisthocotylea was grouped as the sister-group of Trematoda (and Cestoda with Monopisthocotylea), where it is also possible that the hypothetical ancestor of this clade could have had a diet different from blood. Whilst the adaptive changes involved in feeding were likely important in driving the evolution of Neodermata, better ancestral reconstructions based on diet are required (e.g., inclusion of genomic data from other important groups, such as Aspidosgastrea, although this is currently lacking).

In terms of morphology, there are currently no conclusive morphological synapomor-phies that support this clade. However, Littlewood et al. [2] proposed that Trematoda shares with Polyopisthocotylea a neodermatan type flame bulb accompanied by a protonephridial capillary with septate junction (see characters 14 and 19 in the matrix of morphological char-acters in the appendices of Littlewood et al. [2]). However, although this characteristic is absent in Cestoda, within Neodermata it can also be found in some Monopisthocotylea, so it is not entirely clear if it is a plesiomorphy. Therefore, it is necessary to continue exploring helminth morphology to find possible synapomorphies in this clade that support it.

In the present study, the lack of evidence of a sister-group relationship between Trema-toda and Cestoda agrees with the absence of any known morphological apomorphies for the Cestoda–Trematode clade [6,12], a clade supported by other studies [1,3]. Likewise, the molecular evidence does not always support a sister-group relationship between Cestoda and Trematoda [82]. As mentioned above, it is likely that cestodes and trematodes diverged independently, therefore the absence of apomorphies is to be expected. This independent divergence could explain the difference in life history traits, such as the use of intermediate hosts (molluscs in Trematoda and mainly crustaceans in Cestoda) or feeding strategies [11].

4.4. Peptidases in Neodermata

The similarity of topologies of the phylogenetic trees based on peptidases and BUSCO and OMA OGPs suggest that these proteins may shed some light on the evolutionary relationships of Neodermata. The peptidases and peptidase inhibitors play an important role in the feeding processes of the neodermatans, with parasite-derived peptidases being particularly important in host tissue digestion [24,86]. Indeed, the diversification of these proteins is suggested to have contributed to the success of the parasitic lifestyle [87]. According to our network and HCPC analyses, the family C01 (papain family) was the most important for the grouping of trematodes + polyopisthocotyleans, I02 (Kunitz-BPTI family) for cestodes, and S01 (chymotrypsin family) for monopisthocotyleans. Because only hematophagous species of trematodes and polyopisthocotyleans were included in this analysis, the question remains of whether the C01 family is also important for non-hematophagous species. The expansions of C01 and S01 could be linked with adaptation to a parasitic lifestyle. For example, the serine peptidases, members of the subfamily S01, may be used by the monopisthocotyleans to digest host tissue during feeding [88]. Peptidases of the C01 family are used by trematodes to penetrate the host and migrate to specific organs, with the cathepsins B, F, and L among the most studied [21,27,68]. Similarly, the proteinase inhibitors perform important functions for the survival of the parasite in the host because they are responsible for inhibiting host enzymes or manipulating host immune responses [24,89]. Members of the also expanded I02 inhibitor family have anticoagulant and anti-inflammatory function [24] and include the Kunitz-type inhibitors that in cestodes suppress the proteolytic activity of their host [89,90].

The present study suggests that peptidases/inhibitors with multiple copies (and those secreted by helminths into host tissues) are under positive selection pressure, which could contribute to the expansion of certain families such as C01, C19, I02, and S01 (see Figure 2 and Supplementary Figures S7 and S9). As these participate in the host–parasite interaction [22,23,27], it is likely that natural selection is acting on them. Indeed, Li et al. [91] observed that the secreted proteins of bacteria and fungi evolve faster than non-secretory proteins, which they suggested is due to selective pressure helping to shape proteins with particular biochemical adaptations to the environment.

Members of the C01 (papain) cysteine peptidases, one of the most abundant peptidase families found in the present study, perform functions related to nutrition, tissue invasion, and immune system evasion in helminths [25]. Recent studies have shown that these peptidases, particularly cathepsins B, C, and L, are secreted by the monogenean *E. nippoonicum* [30], with cathepsin L participating in the digestion of host blood proteins [92]. Likewise, the trematode *F. hepatica* uses distinct cathepsin L enzymes to degrade host haemoglobin (FhCL1) and to penetrate the host duodenum/migrate through the liver parenchyma due the unique collagenolytic activity of certain family members (FhCL2 and FhCL3) [27,71,93]. Like other cysteine peptidases, the different substrate preferences of the *Fasciola* cathepsin L family members are conferred by the specific arrangement of amino acids that create the S2 subsite [66,69,70]. For instance, FhCL1 (S2 subsite: Leu^{67}, Val^{157} and Leu^{205}) prefers to cleave leucine and phenylalanine-containing substrates (haemoglobin is particularly rich in these) so it may have specifically evolved to facilitate blood feeding [93]. FhCL2 (S2 subsite: Tyr^{67}, Leu^{157} and Leu^{205}) and FhCL3 (S2 subsite: Trp^{67}, Val^{157} and Val^{205}) both accept proline-containing substrates and can digest collagen, thus aiding in host tissue degradation [66,71,72]. It is noteworthy that the majority of cathepsins L from the varied taxonomic groups studied here displayed an S2 subsite topology similar to FhCL1 (i.e., Leu^{67}, Val^{157}) which could suggest roles for these enzymes in blood feeding. However, FhCL1 has activity beyond haemoglobin degradation (reviewed by Cwiklinski et al. [68]), so this S2 subsite arrangement would conceivably allow cleavage of a variety of host macromolecules depending on host niche and mechanism of nutrient acquisition. Nine cathepsin L sequences displayed an S2 subsite arrangement similar to FhCL2 (i.e., Tyr^{67}). Collagen is a major component of vertebrate connective tissue so the possible collagenase activity of these enzymes would aid parasite migration through host tissues and, in the case of monopisthocotyleans, help degrade the epidermis. In the present study, the phylogenetic relatedness of *F. hepatica* cathepsins L with those of other taxa of Neodermata suggest that these peptidases are important for the adaptation and evolution of neodermatans more generally.

The expansion of the S01C subfamily, mainly in most of the monopisthocotyleans studied here, indicates that these peptidases are particularly important for this group of monogeneans. Members of S01C have essential roles in protein digestion and pathogen invasion [22]. Ingram et al. [67] reported that cercarial elastase in *S. mansoni* (a S01C family member) is essential for host skin penetration and that the expansion of this group of peptidases may imply the acquisition of new functions related to host invasion. Future experimental studies are required to investigate the role of the S01C peptidases identified in this study, during infection by species such as *R. viridisi*, *G. salaris*, *N. melleni*, and *E. nipponicum* [94–96].

5. Concluding Remarks

In summary, based on putative OGPs, this study provides evidence in favour of the monophyly of polyopisthocotyleans + trematodes, which has not been discussed in previous studies. In addition, we detected a possible sister-group relationship between monopisthocotyleans and cestodes, although its statistical support was low. To the best of our knowledge, it is the first study to include peptidases in order to clarify the evolutionary relationships within Neodermata. We observed that multicopy and parasite-secreted peptidases/inhibitors were subject to higher selective pressure than single-copy and non-secreted peptidases/inhibitors, which could explain the expansion of some families such as C01, C19, I02, and S01, involved in host–parasite interaction. Whilst not definitive, we hope that our study will stimulate further research and debate on the evolution of Neodermata.

Supplementary Materials: The following supporting information can be downloaded at: https: //www.mdpi.com/article/10.3390/tropicalmed8010059/s1, Figure S1: Constrained trees constructed of 13 platyhelminth species, based on single-copy BUSCO OGPs and simply OMA OGPs; Figure S2: Number of peptidases (A) and peptidase inhibitors (B) identified in the selected platyhelminth species; Figure S3: Number and percentage of peptidase classes in each flatworm species; Figure S4: Formation of clusters of peptidases/inhibitors OGs and platyhelminth species in network analysis; Figure S5: Distribution of Ka and Ks in peptidases and inhibitors of platyhelminths. (A) Ka was higher in multiple-copy OGs than in single-copy OGs. (B) Ks was similar between single-copy OGs and multiple-copy OGs; Figure S6: Distribution of Ka and Ks in peptidases and inhibitors of platyhelminths, with the center of each legend positioned at the highest peak of the distribution; Figure S7: Phylogenetic analysis of the C01A peptidase subfamily. The alignment of the sequences was used to identify the S2 subsite residues belonging to cathepsins L (position 67, 157, and 205) in platyhelminths; Figure S8: Frequency of the amino acids comprising the S2 subsite (positions 67, 167 and 205) of cathepsins L belonging to the C01A subfamily in platyhelminths; Figure S9: Phylogenetic analysis of the S01C peptidase subfamily; Figure S10: Gene Ontology annotation of proteins used in the phylogenomic analysis; Figure S11: Number of shared peptidase families; Table S1: BUSCO OGPs of 13 platyhelminth species; Table S2: OMA OGPs of 13 platyhelminth species; Table S3: Peptidase and inhibitors identified in 13 platyhelminth species; Table S4: Number and percentage of peptidase classes in each flatworm species; Table S5: Number and percentage of peptidase families in each flatworm species; Table S6: Number of peptidases with predicted signal peptides in the 13 platyhelminth species analysed; Table S7: OMA OGs of peptidases of 13 platyhelminth species; Table S8: OGs of peptidases of 13 platyhelminth species used in phylogenetic analysis; Table S9: Number of OGs of peptidases by family used in phylogenetic analysis; Table S10: OGs of peptidases and inhibitors that most contribute to the formation of groups network analyses; Table S11: OGs of peptidases and inhibitors that most contribute to the formation of groups in the HCPC analyses; Table S12: Annotation of the OGs of peptidases and inhibitors that most contribute to the formation of groups in the HCPC analyses; Table S13: Ka and Ks of peptidases and inhibitors of platyhelminths; Table S14: Members of the C01A and S01C subfamilies that were classified in the phylogenetic analysis; Table S15: The S2 subsite residues belonging to cathepsins L (position 67, 157, and 205) identified in platyhelminths; Table S16: Phylogenetic results of Neodermata in other studies; File S1: Fasta files of the 137 BUSCO OGPs used in the phylogenetic analyses; File S2: Fasta files of the 479 OMA OGPs used in the phylogenetic analyses; File S3: Fasta files of the 3960 peptidases identified in the 13 species of platyhelminths; File S4: Fasta files of the 531 OGs of peptidases and peptidase inhibitors in the 13 species of platyhelminths; File S5: Fasta files of the 85 OGs of peptidases used in the phylogenetic analyses; Supplementary Results: Results about the identification of peptidases and inhibitors in 13 platyhelminth species.

Author Contributions: Conceptualization, V.C.-B. and F.N.M.-S.; methodology, V.C.-B.; formal analysis, V.C.-B.; writing—original draft preparation, V.C.-B. and F.N.M.-S.; writing—review and editing, M.W.R. and D.I.H.-M.; supervision, F.N.M.-S. and M.W.R. funding acquisition, F.N.M.-S. All authors have read and agreed to the published version of the manuscript.

Funding: This research was funded by the National Council of Science and Technology of Mexico (CONACYT) through the grant FORDECYT-PRONACES CF/1715616.

Institutional Review Board Statement: Not applicable.

Informed Consent Statement: Not applicable.

Data Availability Statement: All data files generated in this study are available in the Supplementary Materials.

Acknowledgments: V.C.-B. thanks CONACYT for his graduate studentship.

Conflicts of Interest: The authors declare no conflict of interest.

References

1. Hahn, C.; Fromm, B.; Bachmann, L. Comparative genomics of flatworms (Platyhelminthes) reveals shared genomic features of ecto-and endoparastic neodermata. *Genome Biol. Evol.* **2014**, *6*, 1105–1117. [CrossRef]
2. Littlewood, D.T.J.; Rohde, K.; Clough, K.A. Interrelationships of all major groups of Platyhelminthes: Phylogenetic evidence from morphology and molecules. *Biol. J. Linn. Soc.* **1999**, *66*, 75–114. [CrossRef]
3. Perkins, E.M.; Donnellan, S.C.; Bertozzi, T.; Whittington, I.D. Closing the mitochondrial circle on paraphyly of the Monogenea (Platyhelminthes) infers evolution in the diet of parasitic flatworms. *Int. J. Parasitol.* **2010**, *40*, 1237–1245. [CrossRef]
4. Zhang, D.; Li, W.X.; Zou, H.; Wu, S.G.; Li, M.; Jakovlić, I.; Zhang, J.; Chen, R.; Wang, G. Homoplasy or plesiomorphy? Reconstruction of the evolutionary history of mitochondrial gene order rearrangements in the subphylum Neodermata. *Int. J. Parasitol.* **2019**, *49*, 819–829. [CrossRef]
5. Boeger, W.A.; Kritsky, D.C. Phylogenetic relationships of the Monogenoidea. In *Interrelationships of the Platyhelminthes*; Littlewood, D.T.J., Bray, R.A., Eds.; Taylor and Francis: London, UK, 2001; pp. 92–102.
6. Lockyer, A.E.; Olson, P.D.; Littlewood, D.T.J. Utility of complete large and small subunit rRNA genes in resolving the phylogeny of the Neodermata (Platyhelminthes): Implications and a review of the cercomer theory. *Biol. J. Linn. Soc.* **2003**, *78*, 155–171. [CrossRef]
7. Boeger, W.A.; Kritsky, D.C. Phylogeny and a revised classification of the Monogenoidea Bychowsky 1937 (Platyhelminthes). *Syst. Parasitol.* **1993**, *26*, 1–32. [CrossRef]
8. Justine, J.L. Cladistic study in the Monogenea "Platyhelminthes" based upon a parsimony analysis of spermiogenetic and spermatozoal ultrastructural characters. *Int. J. Parasitol.* **1991**, *21*, 821–838. [CrossRef]
9. Justine, J.L. Non-monophyly of the monogeneans? *Int. J. Parasitol.* **1998**, *28*, 1653–1657. [CrossRef] [PubMed]
10. Laumer, C.E.; Giribet, G. Inclusive taxon sampling suggests a single, stepwise origin of ectolecithality in Platyhelminthes. *Biol. J. Linn. Soc.* **2014**, *111*, 570–588. [CrossRef]
11. Park, J.K.; Kim, K.H.; Kang, S.; Kim, W.; Eom, K.S.; Littlewood, D.T.J. A common origin of complex life cycles in parasitic flatworms: Evidence from the complete mitochondrial genome of *Microcotyle sebastis* (Monogenea: Platyhelminthes). *BMC Evol. Biol.* **2007**, *7*, 1–13. [CrossRef]
12. Laumer, C.E.; Hejnol, A.; Giribet, G. Nuclear genomic signals of the 'microturbellarian' roots of platyhelminth evolutionary innovation. *eLife* **2015**, *4*, e05503. [CrossRef] [PubMed]
13. Bernt, M.; Bleidorn, C.; Braband, A.; Dambach, J.; Donath, A.; Fritzsch, G.; Golombek, A.; Hadrys, H.; Jühling, F.; Meusemann, K.; et al. A comprehensive analysis of bilaterian mitochondrial genomes and phylogeny. *Mol. Biol. Evol.* **2013**, *69*, 352–364. [CrossRef] [PubMed]
14. Charrier, N.P.; Hermouet, A.; Hervet, C.; Agoulon, A.; Barker, S.C.; Heylen, D.; Toty, C.; McCoy, K.D.; Plantard, O.; Rispe, C. A transcriptome-based phylogenetic study of hard ticks (Ixodidae). *Sci. Rep.* **2019**, *9*, 12923. [CrossRef]
15. Hawkins, J.A.; Kaczmarek, M.E.; Müller, M.A.; Drosten, C.; Press, W.H.; Sawyer, S.L. A metaanalysis of bat phylogenetics and positive selection based on genomes and transcriptomes from 18 species. *Proc. Natl Acad. Sci. USA* **2019**, *116*, 11351–11360. [CrossRef] [PubMed]
16. Gadagkar, S.R.; Rosenberg, M.S.; Kumar, S. Inferring species phylogenies from multiple genes: Concatenated sequence tree versus consensus gene tree. *J. Exp. Zool. B Mol. Dev. Evol.* **2005**, *304*, 64–74. [CrossRef]
17. Philippe, H.; de Vienne, D.M.; Ranwez, V.; Roure, B.; Baurain, D.; Delsuc, F. Pitfalls in supermatrix phylogenomics. *Eur. J. Taxon.* **2017**, *283*, 1–25. [CrossRef]
18. Waterhouse, R.M.; Seppey, M.; Simão, F.A.; Manni, M.; Ioannidis, P.; Klioutchnikov, G.; Kriventseva, E.V.; Zdobnov, E.M. BUSCO applications from quality assessments to gene prediction and phylogenomics. *Mol. Biol. Evol.* **2018**, *35*, 543–548. [CrossRef]
19. Dylus, D.; Nevers, Y.; Altenhoff, A.M.; Gürtler, A.; Dessimoz, C.; Glover, N.M. How to build phylogenetic species trees with OMA. *F1000Research* **2020**, *9*, 511. [CrossRef]

20. Patwardhan, A.; Ray, S.; Roy, A. Molecular markers in phylogenetic studies-a review. *J. Phylogen. Evol. Biol.* **2014**, *2*, 131. [CrossRef]

21. Kasný, M.; Mikes, L.; Hampl, V.; Dvorak, J.; Caffrey, C.R.; Dalton, J.P.; Horak, P. Peptidases of trematodes. *Adv. Parasitol.* **2009**, *69*, 205–297. [CrossRef]

22. Dvorak, J.; Horn, M. Serine proteases in schistosomes and other trematodes. *Int. J. Parasitol.* **2018**, *48*, 333–344. [CrossRef] [PubMed]

23. Jedličková, L.; Dvořáková, H.; Dvořák, J.; Kašný, M.; Ulrychová, L.; Vorel, J.; Žárský, V.; Mikeš, L. Cysteine peptidases of *Eudiplozoon nipponicum*: A broad repertoire of structurally assorted cathepsins L in contrast to the scarcity of cathepsins B in an invasive species of haematophagous monogenean of common carp. *Parasit. Vectors* **2018**, *11*, 142. [CrossRef] [PubMed]

24. Ranasinghe, S.L.; McManus, D.P. Protease inhibitors of parasitic flukes: Emerging roles in parasite survival and immune defence. *Trends Parasitol.* **2017**, *33*, 400–413. [CrossRef] [PubMed]

25. Caffrey, C.R.; Goupil, L.; Rebello, K.M.; Dalton, J.P.; Smith, D. Cysteine proteases as digestive enzymes in parasitic helminths. *PLoS Negl. Trop. Dis.* **2018**, *12*, e0005840. [CrossRef] [PubMed]

26. Zhang, J. Evolution by gene duplication: An update. *Trends Ecol. Evol.* **2003**, *18*, 292–298. [CrossRef]

27. Robinson, M.W.; Dalton, J.P.; Donnelly, S. Helminth pathogen cathepsin proteases: It's a family affair. *Trends Biochem. Sci.* **2008**, *33*, 601–608. [CrossRef]

28. Caña-Bozada, V.; Morales-Serna, F.N.; Fajer-Ávila, E.J.; Llera-Herrera, R. De novo transcriptome assembly and identification of GPCRs in two species of monogenean parasites of fish. *Parasite* **2022**, *29*, 51. [CrossRef]

29. Howe, K.L.; Bolt, B.J.; Shafie, M.; Kersey, P.; Berriman, M. WormBase ParaSite—A comprehensive resource for helminth genomics. *Mol. Biochem. Parasitol.* **2017**, *215*, 2–10. [CrossRef]

30. Vorel, J.; Cwiklinski, K.; Roudnický, P.; Ilgová, J.; Jedličková, L.; Dalton, J.P.; Mikeš, L.; Gelnar, M.; Kašný, M. *Eudiplozoon nipponicum* (Monogenea, Diplozoidae) and its adaptation to haematophagy as revealed by transcriptome and secretome profiling. *BMC Genom.* **2021**, *22*, 274. [CrossRef]

31. Seppey, M.; Manni, M.; Zdobnov, E.M. BUSCO: Assessing genome assembly and annotation completeness. In *Gene Prediction. Methods in Molecular Biology*; Kollmar, M., Ed.; Humana: New York, NY, USA, 2019; Volume 1962, pp. 227–245. [CrossRef]

32. Altenhoff, A.M.; Levy, J.; Zarowiecki, M.; Tomiczek, B.; Vesztrocy, A.W.; Dalquen, D.A.; Müller, S.; Telford, M.J.; Glover, N.M.; Dylus, D.; et al. OMA standalone: Orthology inference among public and custom genomes and transcriptomes. *Genome Res.* **2019**, *29*, 1152–1163. [CrossRef]

33. Kriventseva, E.V.; Kuznetsov, D.; Tegenfeldt, F.; Manni, M.; Dias, R.; Simão, F.A.; Zdobnov, E.M. OrthoDB v10: Sampling the diversity of animal, plant, fungal, protist, bacterial and viral genomes for evolutionary and functional annotations of orthologs. *Nucleic Acids Res.* **2019**, *47*, D807–D811. [CrossRef] [PubMed]

34. Camacho, C.; Coulouris, G.; Avagyan, V.; Ma, N.; Papadopoulos, J.; Bealer, K.; Madden, T.L. BLAST+: Architecture and applications. *BMC Bioinform.* **2009**, *10*, 421. [CrossRef] [PubMed]

35. Törönen, P.; Medlar, A.; Holm, L. PANNZER2: A rapid functional annotation web server. *Nucleic Acids Res.* **2018**, *46*, W84–W88. [CrossRef] [PubMed]

36. Ye, J.; Zhang, Y.; Cui, H.; Liu, J.; Wu, Y.; Cheng, Y.; Xu, H.; Huang, X.; Li, S.; Zhou, A.; et al. WEGO 2.0: A web tool for analyzing and plotting GO annotations, 2018 update. *Nucleic Acids Res.* **2018**, *46*, W71–W75. [CrossRef] [PubMed]

37. Edgar, R.C. MUSCLE: Multiple sequence alignment with high accuracy and high throughput. *Nucleic Acids Res.* **2004**, *32*, 1792–1797. [CrossRef] [PubMed]

38. Capella-Gutiérrez, S.; Silla-Martínez, J.M.; Gabaldón, T. trimAl: A tool for automated alignment trimming in large-scale phylogenetic analyses. *Bioinformatics* **2009**, *25*, 1972–1973. [CrossRef]

39. Kalyaanamoorthy, S.; Minh, B.Q.; Wong, T.K.; Von Haeseler, A.; Jermiin, L.S. ModelFinder: Fast model selection for accurate phylogenetic estimates. *Nat. Methods* **2017**, *14*, 587–589. [CrossRef]

40. Nguyen, L.T.; Schmidt, H.A.; Von Haeseler, A.; Minh, B.Q. IQ-TREE: A fast and effective stochastic algorithm for estimating maximum-likelihood phylogenies. *Mol. Biol. Evol.* **2015**, *32*, 268–274. [CrossRef]

41. Stamatakis, A. RAxML version 8: A tool for phylogenetic analysis and post-analysis of large phylogenies. *Bioinformatics* **2014**, *30*, 1312–1313. [CrossRef]

42. Campos, A.; Cummings, M.P.; Reyes, J.L.; Laclette, J.P. Phylogenetic relationships of Platyhelminthes based on 18S ribosomal gene sequences. *Mol. Biol. Evol.* **1998**, *10*, 1–10. [CrossRef]

43. Kishino, H.; Miyata, T.; Hasegawa, M. Maximum likelihood inference of protein phylogeny and the origin of chloroplasts. *J. Mol. Evol.* **1990**, *31*, 151–160. [CrossRef]

44. Kishino, H.; Hasegawa, M. Evaluation of the maximum likelihood estimate of the evolutionary tree topologies from DNA sequence data, and the branching order in hominoidea. *J. Mol. Evol.* **1989**, *29*, 170–179. [CrossRef] [PubMed]

45. Shimodaira, H.; Hasegawa, M. Multiple comparisons of log-likelihoods with applications to phylogenetic inference. *Mol. Biol. Evol.* **1999**, *16*, 1114–1116. [CrossRef]

46. Strimmer, K.; Rambaut, A. Inferring confidence sets of possibly misspecified gene trees. *Proc. Biol. Sci.* **2002**, *269*, 137–142. [CrossRef] [PubMed]

47. Shimodaira, H. An Approximately Unbiased Test of Phylogenetic Tree Selection. *Syst. Biol.* **2002**, *51*, 492–508. [CrossRef]

48. Rawlings, N.D.; Barrett, A.J.; Bateman, A. MEROPS: The peptidase database. *Nucleic Acids Res.* **2010**, *38*, D227–D233. [CrossRef]

49. Fu, L.; Niu, B.; Zhu, Z.; Wu, S.; Li, W. CD-HIT: Accelerated for clustering the next-generation sequencing data. *Bioinformatics* **2012**, *28*, 3150–3152. [CrossRef]
50. Ji, Y.; Zhang, Z.; Hu, Y. The repertoire of G-protein-coupled receptors in *Xenopus tropicalis*. *BMC Genom.* **2009**, *10*, 263. [CrossRef]
51. Petersen, T.N.; Brunak, S.; Von Heijne, G.; Nielsen, H. SignalP 4.0: Discriminating signal peptides from transmembrane regions. *Nat. Methods* **2011**, *8*, 785–786. [CrossRef]
52. Bendtsen, J.D.; Jensen, L.J.; Blom, N.; Von Heijne, G.; Brunak, S. Feature-based prediction of non-classical and leaderless protein secretion. *Protein Eng. Des. Sel.* **2004**, *17*, 349–356. [CrossRef]
53. Bastian, M.; Heymann, S.; Jacomy, M. Gephi: An open source software for exploring and manipulating networks. In Proceedings of the International AAAI Conference on Web and Social Media, Atlanta, GA, USA, 6–9 June 2009; Volume 3, pp. 361–362.
54. Blondel, V.D.; Guillaume, J.L.; Lambiotte, R.; Lefebvre, E. Fast unfolding of communities in large networks. *J. Stat. Mech. Theory Exp.* **2008**, *2008*, P10008. [CrossRef]
55. Fortunato, S. Community detection in graphs. *Phys. Rep.* **2010**, *486*, 75–174. [CrossRef]
56. Lázár, A.; Abel, D.; Vicsek, T. Modularity measure of networks with overlapping communities. *Europhys. Lett.* **2010**, *90*, 18001. [CrossRef]
57. Husson, F.; Josse, J.; Le, S.; Mazet, J.; Husson, M.F. Package 'FactoMineR'. *R Package* **2016**, *96*, 698.
58. Odong, T.L.; Van Heerwaarden, J.; van Hintum, T.J.L.; van Eeuwijk, F.A.; Jansen, J. Improving hierarchical clustering of genotypic data via principal component analysis. *Crop Sci.* **2013**, *53*, 1546–1554. [CrossRef]
59. Gu, Z.; Gu, L.; Eils, R.; Schlesner, M.; Brors, B. *Circlize* implements and enhances circular visualization in R. *Bioinformatics* **2014**, *30*, 2811–2812. [CrossRef]
60. Katoh, K.; Standley, D.M. MAFFT multiple sequence alignment software version 7: Improvements in performance and usability. *Mol. Biol. Evol.* **2013**, *30*, 772–780. [CrossRef]
61. Zhang, Z.; Xiao, J.; Wu, J.; Zhang, H.; Liu, G.; Wang, X.; Dai, L. ParaAT: A parallel tool for constructing multiple protein-coding DNA alignments. *Biochem. Biophys. Res. Commun.* **2012**, *419*, 779–781. [CrossRef]
62. Yang, Z. PAML 4: Phylogenetic analysis by maximum likelihood. *Mol. Biol. Evol.* **2007**, *24*, 1586–1591. [CrossRef]
63. Caña-Bozada, V.; Morales-Serna, F.N. PANAS: Pipeline and a case study to obtain synonymous and nonsynonymous substitution rates in genes of Platyhelminthes. *Comp. Parasitol.* 2023; *in press*.
64. Wang, S.; Wang, S.; Luo, Y.; Xiao, L.; Luo, X.; Gao, S.; Dou, Y.; Zhang, H.; Guo, A.; Meng, Q.; et al. Comparative genomics reveals adaptive evolution of Asian tapeworm in switching to a new intermediate host. *Nat. Commun.* **2016**, *7*, 12845. [CrossRef]
65. Letunic, I.; Bork, P. Interactive Tree Of Life (iTOL) v5: An online tool for phylogenetic tree display and annotation. *Nucleic Acids Res.* **2021**, *49*, W293–W296. [CrossRef]
66. Stack, C.M.; Caffrey, C.R.; Donnelly, S.M.; Seshaadri, A.; Lowther, J.; Tort, J.F.; Collins, P.R.; Robinson, M.W.; Xu, W.; McKerrow, J.H.; et al. Structural and functional relationships in the virulence-associated cathepsin L proteases of the parasitic liver fluke, *Fasciola hepatica*. *J. Biol. Chem.* **2008**, *283*, 9896–9908. [CrossRef]
67. Ingram, J.R.; Rafi, S.B.; Eroy-Reveles, A.A.; Ray, M.; Lambeth, L.; Hsieh, I.; Ruelas, D.; Lim, K.C.; Sakanari, J.; Craik, C.S.; et al. Investigation of the proteolytic functions of an expanded cercarial elastase gene family in *Schistosoma mansoni*. *PLoS Negl. Trop. Dis.* **2012**, *6*, e1589. [CrossRef]
68. Cwiklinski, K.; Donnelly, S.; Drysdale, O.; Jewhurst, H.; Smith, D.; Verissimo, C.D.M.; Pritsch, I.C.; O'Neill, S.; Dalton, J.P.; Robinson, M.W. The cathepsin-like cysteine peptidases of trematodes of the genus *Fasciola*. *Adv. Parasitol.* **2019**, *104*, 113–164. [CrossRef]
69. Turk, D.; Gunčar, G.; Podobnik, M.; Turk, B. Revised definition of substrate binding sites of papain-like cysteine proteases. *Biol. Chem.* **1998**, *379*, 137–147. [CrossRef]
70. Lecaille, F.; Choe, Y.; Brandt, W.; Li, Z.; Craik, C.S.; Brömme, D. Selective inhibition of the collagenolytic activity of human cathepsin K by altering its S2 subsite specificity. *Biochemistry* **2002**, *41*, 8447–8454. [CrossRef]
71. Corvo, I.; Cancela, M.; Cappetta, M.; Pi-Denis, N.; Tort, J.F.; Roche, L. The major cathepsin L secreted by the invasive juvenile *Fasciola hepatica* prefers proline in the S2 subsite and can cleave collagen. *Mol. Biochem. Parasitol.* **2009**, *167*, 41–47. [CrossRef]
72. Robinson, M.W.; Corvo, I.; Jones, P.M.; George, A.M.; Padula, M.P.; To, J.; Cancela, M.; Rinaldi, G.; Tort, J.F.; Roche, L.; et al. Collagenolytic activities of the major secreted cathepsin L peptidases involved in the virulence of the helminth pathogen, *Fasciola hepatica*. *PLoS Negl. Trop. Dis.* **2011**, *5*, e1012. [CrossRef]
73. Lambert, A. Oncomiracidiums et phylogenèse des Monogenea (Plathelminthes)-lre Partie: Développement post-larvaire. *Ann. Parasitol. Hum. Comp.* **1980**, *55*, 165–198. [CrossRef]
74. Brooks, D.R.; O'Grady, R.T.; Glen, D.R. The phylogeny of the cercomeria Brooks, 1982 (Platyhelminthes). *Proc. Helminthol. Soc. Wash.* **1985**, *52*, 606–616.
75. Ehlers, U. Comments on a phylogenetic system of the Platyhelminthes. In *Advances in the Biology of Turbellarians and Related Platyhelminthes. Developments in Hydrobiology*; Tyler, S., Ed.; Springer: Berlin/Heidelberg, Germany, 1986; Volume 32. [CrossRef]
76. Rohde, K. Phylogeny of Platyhelminthes, with special reference to parasitic groups. *Int. J. Parasitol.* **1990**, *20*, 979–1007. [CrossRef]
77. Justine, J.L.; Lambert, A.; Mattei, X. Spermatozoon ultrastructure and phylogenetic relationships in the monogeneans (Platyhelminthes). *Int. J. Parasitol.* **1985**, *15*, 601–608. [CrossRef]
78. Baverstock, P.R.; Fielke, R.; Johnson, A.M.; Bray, R.A. Beveridge I. Conflicting phylogenetic hypotheses for the parasitic platyhelminths tested by partial sequencing of 18S ribosomal RNA. *Int. J. Parasitol.* **1991**, *21*, 329–339. [CrossRef] [PubMed]

79. Rohde, K.; Hefford, C.; Ellis, J.T.; Baverstock, P.R.; Johnson, A.M.; Watson, N.A.; Dittmann, S. Contributions to the phylogeny of Platyhelminthes based on partial sequencing of 18S ribosomal DNA. *Int. J. Parasitol.* **1993**, *23*, 705–724. [CrossRef]

80. Mollaret, I.; Jamieson, B.G.; Adlard, R.D.; Hugall, A.; Lecointre, G.; Chombard, C.; Justine, J.L. Phylogenetic analysis of the Monogenea and their relationships with Digenea and Eucestoda inferred from 28S rDNA sequences. *Mol. Biochem. Parasitol.* **1997**, *90*, 433–438. [CrossRef]

81. Litvaitis, M.K.; Rohde, K. A molecular test of platyhelminth phylogeny: Inferences from partial 28s rDNA sequences. *Invertebr. Biol.* **1999**, *118*, 42–56. [CrossRef]

82. International Helminth Genomes Consortium. Comparative genomics of the major parasitic worms. *Nat. Genet.* **2019**, *51*, 163–174. [CrossRef]

83. Littlewood, D.T.J.; Rohde, K.; Bray, R.A.; Herniou, E.A. Phylogeny of the Platyhelminthes and the evolution of parasitism. *Biol. J. Linn. Soc.* **1999**, *68*, 257–287. [CrossRef]

84. Bychowsky, B.E. Ontogenesis and phylogenetic interrelations of parasitic flatworms. *Izvest. Akad. Nauk SSSR Ser. Biol.* **1937**, *4*, 1353–1383.

85. Xylander, W.E.R. The Gyrocotylidea, Amphilinidea and the early evolution of Cestoda. In *Interrelationships of the Platyhelminthes*; Littlewood, D.T.J., Bray, R.A., Eds.; Taylor and Francis, Inc.: New York, NY, USA, 2001; pp. 103–111.

86. Jedličková, L.; Dvořák, J.; Hrachovinová, I.; Ulrychová, L.; Kašný, M.; Mikeš, L. A novel Kunitz protein with proposed dual function from *Eudiplozoon nipponicum* (Monogenea) impairs haemostasis and action of complement in vitro. *Int. J. Parasitol.* **2019**, *49*, 337–346. [CrossRef]

87. Dalton, J.P.; Skelly, P.; Halton, D.W. Role of the tegument and gut in nutrient uptake by parasitic platyhelminths. *Can. J. Zool.* **2004**, *82*, 211–232. [CrossRef]

88. Hirazawa, N.; Umeda, N.; Hatanaka, A.; Kuroda, A. Characterization of serine proteases in the monogenean *Neobenedenia girellae*. *Aquaculture* **2006**, *255*, 188–195. [CrossRef]

89. Rogozhin, E.A.; Solovyev, M.M.; Frolova, T.V.; Izvekova, G.I. Isolation and partial structural characterization of new Kunitz-type trypsin inhibitors from the pike cestode *Triaenophorus nodulosus*. *Mol. Biochem. Parasitol.* **2019**, *233*, 111217. [CrossRef] [PubMed]

90. Izvekova, G.I.; Frolova, T.V.; Izvekov, E.I. Adsorption and inactivation of proteolytic enzymes by *Triaenophorus nodulosus* (Cestoda). *Helminthologia* **2017**, *54*, 3–10. [CrossRef]

91. Li, Y.D.; Xie, Z.Y.; Du, Y.L.; Zhou, Z.; Mao, X.M.; Lv, L.X.; Li, Y.Q. The rapid evolution of signal peptides is mainly caused by relaxed selection on non-synonymous and synonymous sites. *Gene* **2009**, *436*, 8–11. [CrossRef]

92. Jedličková, L.; Dvořáková, H.; Kašný, M.; Ilgová, J.; Potěšil, D.; Zdráhal, Z.; Mikeš, L. Major acid endopeptidases of the blood-feeding monogenean *Eudiplozoon nipponicum* (Heteronchoinea: Diplozoidae). *Parasitology* **2016**, *143*, 494–506. [CrossRef]

93. Lowther, J.; Robinson, M.W.; Donnelly, S.M.; Xu, W.; Stack, C.M.; Matthews, J.M.; Dalton, J.P. The importance of pH in regulating the function of the *Fasciola hepatica* cathepsin L1 cysteine protease. *PLoS Negl. Trop. Dis.* **2009**, *3*, e369. [CrossRef]

94. Kawatsu, H. Studies on the anemia of fish–IX. Hypochromic microcytic anemia of crucian carp caused by infestation with a trematode, *Diplozoon nipponicum*. *Bull. Japan. Soc. Sci. Fish.* **1978**, *44*, 1315–1319. [CrossRef]

95. Thoney, D.A.; Hargis Jr, W.J. Monogenea (Platyhelminthes) as hazards for fish in confinement. *Annu. Rev. Fish Dis.* **1991**, *1*, 133–153. [CrossRef]

96. Morales-Serna, F.N.; Martínez-Brown, J.M.; Avalos-Soriano, A.; Sarmiento-Vázquez, S.; Hernández-Inda, Z.L.; Medina-Guerrero, R.M.; Fajer-Ávila, E.J.; Ibarra-Castro, L. The efficacy of Geraniol and ß-Citronellol against freshwater and marine monogeneans. *J. Aquati. Anim. Health* **2020**, *32*, 127–132. [CrossRef]

Tropical Medicine and Infectious Disease

Article

Identification of Hazard and Socio-Demographic Patterns of Dengue Infections in a Colombian Subtropical Region from 2015 to 2020: Cox Regression Models and Statistical Analysis

Santiago Ortiz [1], Alexandra Catano-Lopez [1], Henry Velasco [1], Juan P. Restrepo [1], Andrés Pérez-Coronado [1], Henry Laniado [1] and Víctor Leiva [2,*]

[1]   School of Applied Sciences and Engineering, Universidad EAFIT, Medellín 050022, Colombia
[2]   School of Industrial Engineering, Pontificia Universidad Católica de Valparaíso, Valparaíso 2362807, Chile
*   Correspondence: victor.leiva@pucv.cl or victorleivasanchez@gmail.com

Abstract: Dengue is a disease of high interest for public health in the affected localities. Dengue virus is transmitted by *Aedes* species and presents hyperendemic behaviors in tropical and subtropical regions. Colombia is one of the countries most affected by the dengue virus in the Americas. Its central-west region is a hot spot in dengue transmission, especially the Department of Antioquia, which has suffered from multiple dengue outbreaks in recent years (2015–2016 and 2019–2020). In this article, we perform a retrospective analysis of the confirmed dengue cases in Antioquia, discriminating by both subregions and dengue severity from 2015 to 2020. First, we conduct an exploratory analysis of the epidemic data, and then a statistical survival analysis is carried out using a Cox regression model. Our findings allow the identification of the hazard and socio-demographic patterns of dengue infections in the Colombian subtropical region of Antioquia from 2015 to 2020.

Keywords: arbovirus; clinical deterioration; endemic; proportional hazard; tropical diseases

Citation: Ortiz, S.; Catano-Lopez, A.; Velasco, H.; Restrepo, J.P.; Pérez-Coronado, A.; Laniado, H.; Leiva, V. Identification of Hazard and Socio-Demographic Patterns of Dengue Infections in a Colombian Subtropical Region from 2015 to 2020: Cox Regression Models and Statistical Analysis . *Trop. Med. Infect. Dis.* **2023**, *8*, 30. https://doi.org/10.3390/tropicalmed8010030

Academic Editors: Vyacheslav Yurchenko and John Frean

Received: 11 November 2022
Revised: 10 December 2022
Accepted: 21 December 2022
Published: 30 December 2022

1. Introduction

Dengue is a virus transmitted by *Aedes* species and distributed in tropical and subtropical regions [1]. Its main symptoms are fever, headache, and joint pain [2]. Diagnosing infected individuals in regions with the co-circulation of multiple arboviruses, such as zika and chikungunya, is challenging. In the case of dengue, it presents hyperendemic behaviors in tropical and subtropical regions, reporting approximately 96 million clinical cases per year [3]. Dengue is considered one of the most transmitted arboviruses in these regions [4], being endemic in some zones, for example, in the Americas [5].

Socio-economic conditions, changes in dengue over time, and the seasonal temperature variation induced by the El Niño Southern Oscitation (ENSO) influence the epidemiology of this infectious disease. Those aspects make the Americas a suitable place for epidemic outbreaks of dengue. Based on the reported cases, countries such as Brazil, Colombia, and Mexico are the most affected [6]. In the case of Colombia, the severe dengue fatality rate increased over time, as well as its incidence, due to the growing human population, the poor housing infrastructure, environmental drought/high rainfalls, and barriers to accessing health services [6,7].

Central-west Colombia is a dengue transmission hot spot, especially the Department of Antioquia, whose locality has the highest mortality and morbidity in dengue cases [8]. This subtropical locality has suffered from multiple outbreaks in recent years (2015–2016 and 2019–2020) [6,9–11]. Therefore, health authorities there have taken actions such as releasing mosquitoes with Wolbachia, controlling by fumigation, and cleaning breeding sites [12,13].

Dengue is considered a disease of high interest to public health. Several researchers have studied its behavior in both the Americas and Colombia specifically to understand the characteristics of infection and its appearance in the localities of interest [5,7,9,11]. For that reason, some studies implement survival analyses to identify the death or recovery time of a specific disease for vector-borne diseases [14–16]. However, there is still no evidence in the literature of sociodemographic or survival analyses of the impact of dengue on neglected localities in Colombia, such as the Department of Antioquia. Therefore, our main objective is to perform a retrospective analysis of the confirmed cases of dengue in Antioquia, discriminated by subregions and severity type from 2015 to 2020, through a descriptive analysis of the epidemic information and survival analysis using a Cox regression model.

We divide the manuscript as follows. In Section 2, we describe the implemented methodology with an overview of the case of study, the data management methods, the statistical analyses, and the implementation of the robust proportional hazard model. Section 3 reports the main results obtained for the Antioquia case study regarding both subregional behavior and dengue type. In Section 4, the discussion of our findings is presented. Finally, in Section 5, we present some conclusions of this research.

2. Methodology

2.1. Case Study

Antioquia is a Department located in Northwest Colombia (Figure 1a) and has a population of around 6,550,206 inhabitants in an area of 63,612 km^2. Its capital city is the municipality of Medellín (Table 1). Antioquia has entrances to the Caribbean Sea, which is crossed by the central and oriental cordillera. This geographical location generates heterogeneous environments and climates. The humid and semi-humid climates are found in the western, eastern, and central narrow strips. Its temperate and cold climates are distributed in the center of the department along the slopes of the central and western sides. Its precipitation ranges between 1500 mm and 4000 mm per year.

(a)   (b)

Figure 1. Study zone. (**a**) Location of Antioquia Department. (**b**) The territorial subdivision of Antioquia consists of nine subregions that have a high geographic diversity; see Table 1 for further information about each subregion.

Table 1. Summary of geographical and climatic characteristics for the nine subregions in Antioquia. For further information about description of each subregion, see [17,18].

Subregion	Location	Area (km^2)	Altitude (MSL)	Temperature Range (°C)
Valle de Aburrá	South center	1158	1300–1775	12–21
Bajo Cauca	Northeast, in the spur of the CC	8585	30–125	17–29
Norte	North, in CC	7516	1200–2550	12–23
Nordeste	Eastern slopes of the CC	8645	650–1975	19–27
Suroeste	Southwestern, between WC and CC	6589	600–2350	12–26
Occidente	Northwest, between WC and CC	6571	450–1925	10–26
Oriente	Southeast	7103	1000–2500	13–23
Urabá	North, Coastal region	11,799	2–200	22–29
Magdalena Medio	CC	4833	75–950	24–29

Where central cordillera (CC) and west cordillera (WC).

Regarding administrative divisions, Antioquia is distributed by nine subregions with their own environmental, geographical, and social characteristics (Figure 1b and Tables 1 and 2), where 78.8% of its population live in urban zones and 21.2% in rural zones [19]. There are 149,535 registered immigrants, where the maximum number is located in Valle de Aburrá with 122,619, followed by Oriente with 13,458, and Urabá with 4489: nearly 1.9%, 0.2%, and 0.07% of these subregions' populations, respectively. The remaining 8972 immigrants are distributed throughout other subregions [20].

Table 2. Distribution of the population that corresponds to the demographic groups in each subregion [20–24]. * There are 1,307,620 victims reported in Antioquia (nearly 20% of the population); even so, 704,811 of them are not georeferenced [21].

Subregion	Population	Gender Men	Social Groups			Settlement (Urban)	Minorities	
			Disabled	Displaced	Victims *		Indigenous	Mixed-Race and Afro-Colombian
Valle de Aburrá	3,969,222	53%	2.1%	0.07%	2%	97%	0.1%	1.9%
Bajo Cauca	255,064	50%	1.7%	2.21%	21%	65%	2.3%	6.9%
Norte	244,995	51%	3.1%	1.61%	19%	50%	0.22%	1.26%
Nordeste	199,335	50%	2.9%	1.17%	16%	54%	0.46%	0.91%
Suroeste	367,467	50%	3.1%	0.84%	14%	48%	1.22%	0.75%
Occidente	210,371	51%	3.3%	3.65%	24%	38%	4.15%	1.50%
Oriente	683,968	49%	2.6%	2.34%	18%	60%	0.05%	0.37%
Urabá	514,423	49%	1.8%	5.22%	28%	59%	2.67%	39%
Magdalena Medio	105,361	51%	3.7%	2.38%	9%	56%	0.11%	2.55%

The population in Antioquia is divided into the following age groups: early childhood (8.4%), childhood (8.6%), adolescence (10.8%), early adulthood (13.9%), adulthood (44.3%), and old age (14.0%) [20]. In addition, following the population pyramid developed by Gobernación de Antioquia [20], the subregions of Urabá, Magdalena Medio, Bajo Cauca, and Occidente present young populations with a high proportion of children and adolescents and a low number of adults and older adults.

In contrast, the subregions Este, Suroeste, Norte, and Nordeste have shown a decrease in children and adolescents and a gradual widening in the adult population. We highlight that Valle de Aburrá presents a decrease in the birth rate and life expectancy.

Concerning ethnic groups, Colombia recognizes four: (i) indigenous, (ii) Afro-Colombians (includes Afro-descendants, black people, mixed-race people, Palenqueros from San Basilio) (iii) Raizales from the San Andrés and Providencia archipelago, and (iv) ROM (Roma or Gypsy ethnicity). The indigenous groups live in their reservations or in the dispersed rural area of 32 municipalities; even so, this population has gradually mobilized from their territories to large urban centers, presenting a social issue. The ROM, Palenqueros, and Raizales populations are low in Antioquia, registering 140, 183, and 637 people, respectively. Most parts of these minorities are located in Valle de Aburrá and Urabá.

Depending on each subregion, there are localities with favorable conditions for the reproduction and feeding of *Aedes Aegypti*, having temperatures between 24 °C and 31 °C, humidity higher than 60%, and a high presence of breeding sites [25]. The first dengue hemorrhagic fever in Colombia was reported in Antioquia in 1989 [26]. Since then, a dengue alert has been in place throughout the country [8]. The ENSO enhances these arbovirus outbreaks through an increased vector population and feeding habits, as reported in other localities in South America [27,28]. Different outbreaks occurred in Colombia in 2010, 2015–2016, and 2019–2020 [10,11]. According to the Antioquia government, in 2016–2017, the most significant number of cases were in the Valle de Aburrá, Urabá, Occidente, and Suroeste localities. Some subregions could also provide under-registered arbovirus cases since many of them present favorable conditions for mosquito proliferation alongside precarious or non-existent health systems [9,29].

2.2. Dataset

The data to be used correspond to the daily cases of dengue in Antioquia, reported to the Sistema Nacional de Vigilancia en Salud Pública (SIVIGILA, Spanish acronym) during the years 2015–2020 (portalsivigila.ins.gov.co; accessed on 9 December 2020). The SIVIGILA is a system that associates users and procedures for collecting data to be analyzed and to obtain information from these data about health events in Colombia that must be interpreted and distributed. These data come directly from hospitals. The information is reported to the Instituto Nacional de Salud (INS, Spanish acronym), the institution in charge of public health surveillance in Colombia [10]. A case is considered a person who is suspected of infection and is reported to SIVIGILA. The arbovirus cases are classified into five groups: (i) suspect, (ii) probable, (iii) confirmed by the laboratory, (iv) confirmed by the clinic, and (v) confirmed by epidemiological nexus. The similarity between arbovirus symptoms can lead to misclassification biases on probable and suspect cases.

For correct analyses and to avoid misclassification issues, we discard unconfirmed cases, that is, the suspect and probable cases that are not confirmed by the laboratory, the clinic, or by epidemiological nexus [9,10]. Moreover, it is essential to point out that all individuals reported at SIVIGILA were alive patients, as this public institution is not responsible for registering and reporting deaths. We use public domain data collected by official government organizations, freely accessible and available by directly requesting them from these organizations. The health authorities anonymize these data related to social, economic, and symptomatological characteristics for each reported case. These data correspond to age, gender, subregion, municipality, area of occurrence (municipal seat or urban, rural, and dispersed rural), ethnic group (indigenous, ROM, Raizal, Palenquero, mixed-race, and Afro-Colombian), and occupation, according to the International Standard Classification of Occupations (ISCO-08) [30]. In addition, the hospitals report the event dates as epidemiological weeks, symptom onset, hospitalization, and death (the reported deaths correspond only to those directly registered in the SIVIGILA). The cleaned data set and the corresponding computational code in the R statistical software [31] are available at github.com/alexacl95/DengueAntioquia (accessed on 11 July 2022).

2.3. Statistical Analyses

We consider quantitative and qualitative variables according to both subregion location and by type of dengue: common or typical dengue and severe dengue. Subregions only consider socio-demographic variables. For subregional analysis, all quantitative variables are expressed in terms of their sample median, along with their respective 95% bootstrap confidence intervals. For multiple testing, such medians are compared using the Kruskal–Wallis (KW) test. If sample evidence rejects the null hypothesis that the variable differs by subregion, based on the KW test, then all the subregions are compared using a two-sided pairwise Mann–Whitney (MW) test with Holm–Bonferronni correction. Qualitative variables are expressed in their absolute/relative frequencies and compared with subregions by the independence chi-squared test with Yate correction for small samples, if needed.

For comparative analysis by dengue type, quantitative variables are expressed again in terms of their sample median and their respective 95% bootstrap confidence intervals and compared using a two-sided MW test. Similarly, qualitative variables are expressed in their absolute/relative frequencies and compared using a two-sided, two-sample Z test for proportions. We consider the robust proportional hazard model [32], a robust version of the well-known Cox regression [33], to identify covariates that can explain the hazard ratio, avoiding possible interference of outlying observations. This model is formulated for the dependent variable "clinical deterioration time", measured in days, which expresses the delay from symptoms onset to hospital admission. Here, we are interested in measuring the impact of socio-demographic variables on the probability that an infected person requires more days to deteriorate. We perform all statistical analysis in the R software. Specifically the coxrobust package [34] is used to formulate the robust Cox regression model.

3. Results

3.1. Dengue in Antioquia: 2015–2020

In total, 50,397 people with dengue infection were reported, of which only 491 (0.97%) could not be georeferenced. About 80% of the total infections occurred in two subregions, with the majority of cases being located in Valle de Aburrá with 35,335 (70.1%), followed by Urabá with 5196 cases (10.3%). However, the relevant incidence was in Urabá (2017, 2018, and 2020), followed by Magdalena Medio (2015 and 2019), Occidente (2015 and 2017), and Bajo Cauca (2018); see Table 3 and Figure 2. Those subregions are characterized as localities with high temperatures and low altitudes (Table 1): favorable environmental conditions for the reproduction of *Aedes* species.

Table 3. Dengue incidence per 1000 inhabitants by subregions between the years 2015 and 2020.

Subregion	Year						Median Incidence
	2015	**2016**	**2017**	**2018**	**2019**	**2020**	
BC	0.90	0.61	0.57	2.32	2.39	0.55	0.755
MM	1.44	1.11	0.64	0.59	4.00	0.84	0.975
NE	0.64	1.73	0.48	0.13	2.26	0.88	0.760
NO	0.28	0.88	0.12	0.17	0.40	0.08	0.225
OC	1.66	3.51	1.11	0.27	0.88	0.39	0.995
OR	0.20	0.32	0.04	0.03	0.11	0.06	0.085
SO	1.14	4.84	0.50	0.17	0.19	0.63	0.565
UR	1.01	1.01	1.20	2.75	3.08	1.05	1.125
VA	1.29	5.94	0.68	0.38	0.39	0.21	0.535
Median Incidence	1.01	1.11	0.57	0.27	0.88	0.55	0.725

Where Bajo Cauca (BC), Magdalena Medio (MM), Nordeste (NE), Norte (NO), Occidente (OC), Oriente (OR), Suroeste (SO), Urabá (UR), and Valle de Aburrá (VA).

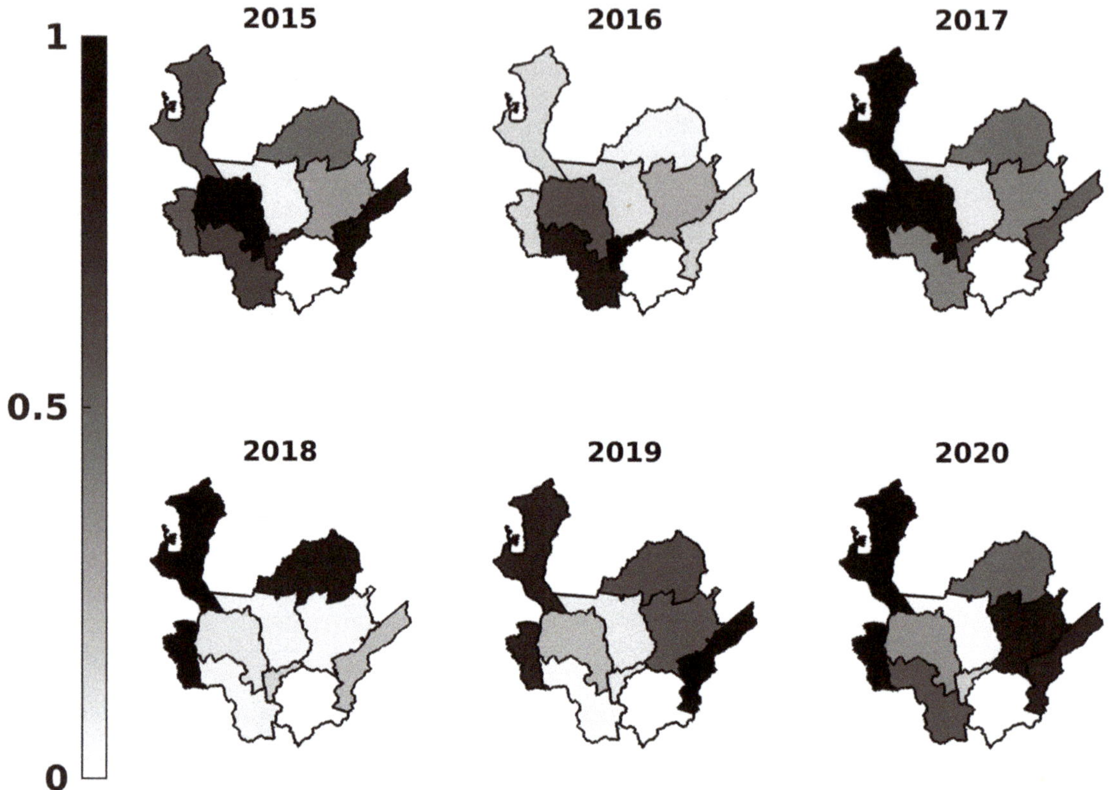

Figure 2. Choropleth map of normalized dengue prevalence in Colombia for cases reported during 2015–2020.

In Figures 2 and 3, we present the number of people affected by dengue over the total population (normalized prevalence per subregion) and the registered dengue cases as time series. A prevalence map is presented using the number of cases per year and the inhabitants per subregion (Table 1), where prevalence in each year was normalized to be between 0 and 1; see Table 3 for the exact values. By normalizing the prevalence values, we can visualize the most affected localities per year. Figure 3 shows the dengue cases' time series per subregion and year corresponding to Southern localities (Suroeste, Occidente, and Valle de Aburra) and Northern localities (Uraba, Bajo Cauca, and Magdalena), which presented outbreaks in 2015–2016 and 2018–2020, respectively. According to Figures 2 and 3, the epidemic outbreaks occurred in different periods for some localities. Suroeste, Occidente, Oriente, and Valle de Aburrá presented outbreaks during 2015–2016, whereas Bajo Cauca, Nordeste, and Magdalena Medio saw increased numbers of endemic cases. The 2015–2016 period corresponds to one of the most intense ENSO phenomena reported in the last years. In addition, Northern localities such as Urabá, Bajo Cauca, Nordeste, and Magdalena Medio suffered dengue outbreaks during 2018–2020 and reported a normal/moderate ENSO effect over the region during this period.

For all socio-demographic variables by subregions, statistical significance was detected using the chi-squared test (Tables 4 and 5). The results indicate that a relationship exists between socio-demographic variables and subregions. Figure 4 summarizes Tables 4 and 5, where we note that men suffer from more dengue infections than women, except in Suroeste and Valle de Aburrá. In addition, people in the adult age group had more infections than other age groups. Nevertheless, in Urabá and Bajo Cauca, younger people (adolescents and children) have a similar prevalence as adults. Regarding the remaining socio-demographic variables, the most affected people, around 80%, were those whose principal occupation was elementary, such as cleaners, agriculture workers, food preparers, mining/construction workers, or street sales workers: occupations that were highly correlated to people with medium-low socioeconomic status. Regarding the location, more than 40% of cases occurred in a municipal capital for all subregions. Hence, according to ethnic minority groups, Afro-Colombians and mixed-race people were the most affected groups in all subregions, especially in Urabá, but in Suroeste and Occidente, indigenous people were the most affected (Figure 4).

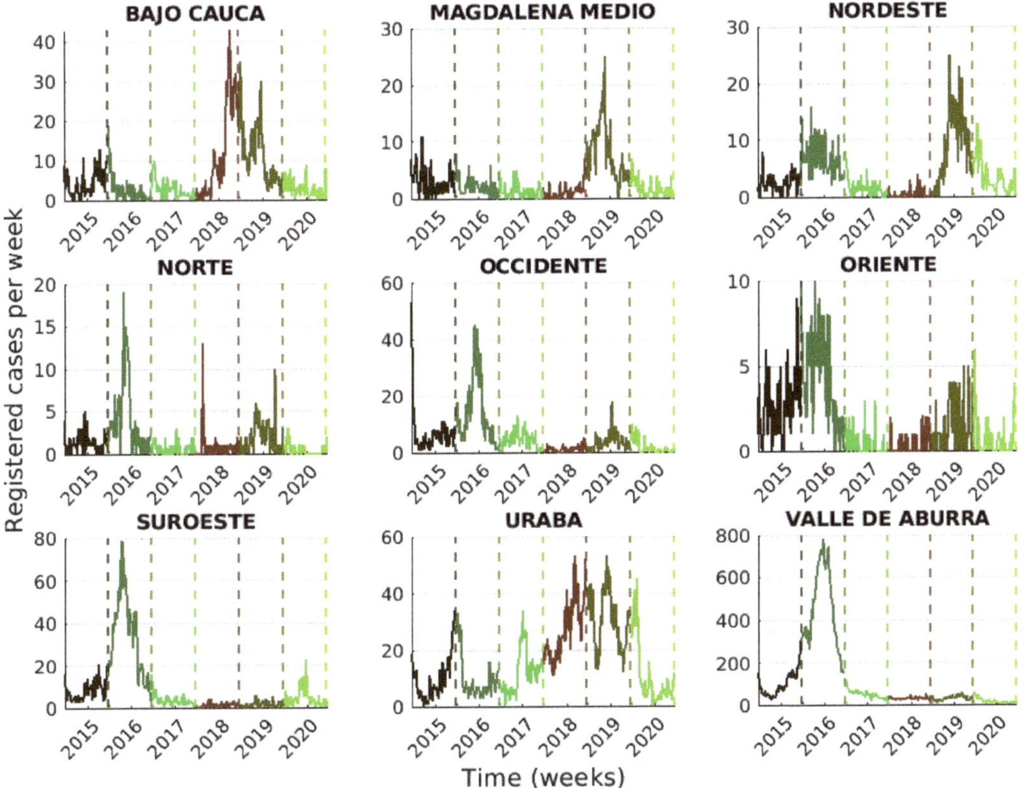

Figure 3. Time-series of reported dengue cases per subregion during 2015–2020. Different colors in the graphs represent the time-frames.

Table 4. Socio-demographic characteristics among patients, according to the subregion.

	Variable	BC (n = 1874)	MM (n = 908)	NE (n = 1218)	NO (n = 475)	OC (n = 1644)	OR (n = 515)	NG (n = 491)	SO (n = 2741)	UR (n = 5196)	VA (n = 35,335)	*p*-Value
	Age	16 (15–16.5)	19 (17–20)	19 (18–21)	28 (26–30)	28.5 (27–29.5)	27 (25–29)	28 (26–31)	30 (29–31)	14 (14–15)	28 (28–29)	<0.0001
Age group	Early childhood (0–5)	241 (12.9%)	89 (9.8%)	149 (12.2%)	20 (4.2%)	90 (5.5%)	38 (7.4%)	37 (7.5%)	95 (3.5%)	943 (18.1%)	2251 (6.4%)	<0.0001
	Childhood (6–11)	426 (22.7%)	150 (16.5%)	229 (18.8%)	42 (8.8%)	125 (7.6%)	35 (6.8%)	46 (9.4%)	234 (8.5%)	1209 (23.3%)	3085 (8.7%)	
	Adolescence (12–18)	399 (21.3%)	213 (23.5%)	200 (16.4%)	65 (13.7%)	241 (14.7%)	90 (17.5%)	62 (12.6%)	446 (16.3%)	1039 (20%)	4773 (13.5%)	
	Early adulthood (19–26)	243 (13%)	129 (14.2%)	204 (16.7%)	98 (20.6%)	301 (18.3%)	90 (17.5%)	85 (17.3%)	444 (16.2%)	622 (12%)	6196 (17.5%)	
	Adulthood (27–59)	468 (25%)	273 (30.1%)	363 (29.8%)	215 (45.3%)	714 (43.4%)	220 (42.7%)	214 (43.6%)	1204 (43.9%)	1154 (22.2%)	15932 (45.1%)	
	Old age (60+)	97 (5.2%)	54 (5.9%)	73 (6%)	35 (7.4%)	173 (10.5%)	42 (8.2%)	47 (9.6%)	318 (11.6%)	229 (4.4%)	3098 (8.8%)	
Sex	Female	820 (43.8%)	398 (43.8%)	548 (45%)	210 (44.2%)	814 (49.5%)	237 (46%)	235 (47.9%)	1392 (50.8%)	2429 (46.7%)	18256 (51.7%)	<0.0001
	Male	1054 (56.2%)	510 (56.2%)	670 (55%)	265 (55.8%)	830 (50.5%)	278 (54%)	256 (52.1%)	1349 (49.2%)	2767 (53.3%)	17079 (48.3%)	
Type of settlement	Municipal capital	1286 (68.6%)	586 (64.5%)	879 (72.2%)	322 (67.8%)	781 (47.5%)	355 (68.9%)	442 (90%)	1849 (67.5%)	2844 (54.7%)	33055 (93.5%)	<0.0001
	Populated center	128 (6.8%)	183 (20.2%)	75 (6.2%)	54 (11.4%)	369 (22.4%)	65 (12.6%)	19 (3.9%)	313 (11.4%)	970 (18.7%)	1443 (4.1%)	
	Rural–dispersed	460 (24.5%)	139 (15.3%)	264 (21.7%)	99 (20.8%)	494 (30%)	95 (18.4%)	30 (6.1%)	579 (21.1%)	1382 (26.6%)	837 (2.4%)	

Table 4. *Cont.*

	Variable	BC (n = 1874)	MM (n = 908)	NE (n = 1218)	NO (n = 475)	OC (n = 1644)	OR (n = 515)	NG (n = 491)	SO (n = 2741)	UR (n = 5196)	VA (n = 35,335)	*p*-Value
Type of occupation (ISCO-08)	Skilled agricultural, forestry, and fishery workers	460 (24.5%)	139 (15.3%)	264 (21.7%)	99 (20.8%)	494 (30%)	95 (18.4%)	30 (6.1%)	579 (21.1%)	1382 (26.6%)	837 (2.4%)	<0.0001
	Managers	5 (0.3%)	8 (0.9%)	2 (0.2%)	1 (0.2%)	13 (0.8%)	4 (0.8%)	9 (1.8%)	23 (0.8%)	14 (0.3%)	328 (0.9%)	
	Armed forces	13 (0.7%)	27 (3%)	10 (0.8%)	4 (0.8%)	6 (0.4%)	8 (1.6%)	11 (2.2%)	6 (0.2%)	29 (0.6%)	63 (0.2%)	
	Elementary occupations	1612 (86%)	751 (82.7%)	953 (78.2%)	339 (71.4%)	1198 (72.9%)	382 (74.2%)	340 (69.2%)	1989 (72.6%)	4398 (84.6%)	27950 (79.1%)	
	Craft and related trades workers	88 (4.7%)	22 (2.4%)	93 (7.6%)	21 (4.4%)	48 (2.9%)	23 (4.5%)	15 (3.1%)	124 (4.5%)	25 (0.5%)	993 (2.8%)	
	Plant and machine operators and assemblers	17 (0.9%)	10 (1.1%)	22 (1.8%)	8 (1.7%)	39 (2.4%)	11 (2.1%)	11 (2.2%)	58 (2.1%)	14 (0.3%)	800 (2.3%)	
	Clerical support workers	8 (0.4%)	8 (0.9%)	6 (0.5%)	5 (1.1%)	37 (2.3%)	9 (1.7%)	9 (1.8%)	42 (1.5%)	21 (0.4%)	840 (2.4%)	
	Professionals	29 (1.5%)	31 (3.4%)	23 (1.9%)	17 (3.6%)	60 (3.6%)	19 (3.7%)	38 (7.7%)	86 (3.1%)	76 (1.5%)	1221 (3.5%)	
	Technicians and associate professionals	12 (0.6%)	15 (1.7%)	17 (1.4%)	13 (2.7%)	51 (3.1%)	16 (3.1%)	30 (6.1%)	56 (2%)	47 (0.9%)	1107 (3.1%)	
	Service and sales workers	31 (1.7%)	21 (2.3%)	34 (2.8%)	25 (5.3%)	85 (5.2%)	26 (5%)	21 (4.3%)	97 (3.5%)	93 (1.8%)	1940 (5.5%)	

Where Bajo Cauca (BC), Magdalena Medio (MM), Nordeste (NE), Norte (NO), Occidente (OC), Oriente (OR), not georeferenced (NG), Suroeste (SO), Urabá (UR), and Valle de Aburrá (VA).

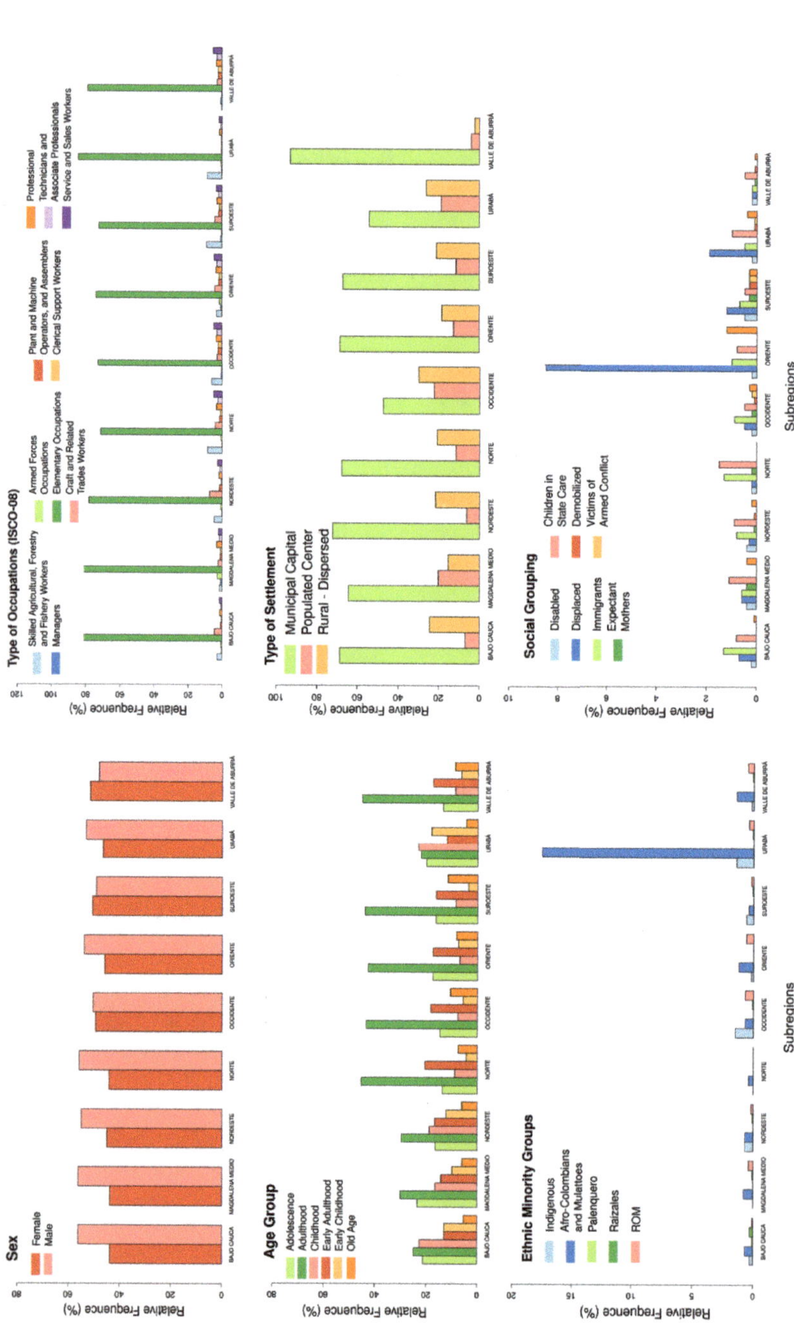

Figure 4. Barcharts for socio-demographic variables, discriminated by subregion, including sex, age groups, ethnic minority groups, type of occupation, and social grouping. For more details and *p*-values, we refer the reader to Tables 4 and 5.

Table 5. Socio-demographic characteristics among patients, according to subregion location (part B).

	Variable	BC (n = 1874)	MM (n = 908)	NE (n = 1218)	NO (n = 475)	OC (n = 1644)	OR (n = 515)	NG (n = 491)	SO (n = 2741)	UR (n = 5196)	VA (n = 35,335)	*p*-Value
Ethnic minority groups	Indigenous	5 (0.3%)	0 (0%)	9 (0.7%)	0 (0%)	24 (1.5%)	1 (0.2%)	2 (0.4%)	16 (0.6%)	72 (1.4%)	66 (0.2%)	<0.0001
	Afro-Colombians and mixed-race	13 (0.7%)	7 (0.8%)	9 (0.7%)	2 (0.4%)	11 (0.7%)	6 (1.2%)	6 (1.2%)	12 (0.4%)	907 (17.5%)	496 (1.4%)	
	Palenquero	1 (0.1%)	0 (0%)	0 (0%)	0 (0%)	0 (0%)	0 (0%)	0 (0%)	0 (0%)	1 (0%)	2 (0%)	
	Raizales	5 (0.3%)	1 (0.1%)	1 (0.1%)	0 (0%)	2 (0.1%)	0 (0%)	1 (0.2%)	2 (0.1%)	4 (0.1%)	28 (0.1%)	
	ROM	2 (0.1%)	4 (0.4%)	3 (0.2%)	0 (0%)	11 (0.7%)	3 (0.6%)	1 (0.2%)	6 (0.2%)	20 (0.4%)	171 (0.5%)	
Social groups	Disabled	4 (0.2%)	4 (0.4%)	5 (0.4%)	1 (0.2%)	4 (0.2%)	1 (0.2%)	3 (0.6%)	15 (0.5%)	12 (0.2%)	59 (0.2%)	0.002
	Displaced	14 (0.7%)	5 (0.6%)	4 (0.3%)	1 (0.2%)	8 (0.5%)	44 (8.5%)	4 (0.8%)	34 (1.2%)	100 (1.9%)	60 (0.2%)	<0.0001
	Immigrants	24 (1.3%)	5 (0.6%)	10 (0.8%)	6 (1.3%)	14 (0.9%)	5 (1%)	4 (0.8%)	20 (0.7%)	25 (0.5%)	70 (0.2%)	<0.0001
	Convicts	0 (0%)	4 (0.4%)	1 (0.1%)	1 (0.2%)	3 (0.2%)	0 (0%)	3 (0.6%)	9 (0.3%)	2 (0%)	27 (0.1%)	<0.0001
	Expectant mothers	15 (0.8%)	10 (1.1%)	11 (0.9%)	7 (1.5%)	8 (0.5%)	4 (0.8%)	2 (0.4%)	15 (0.5%)	51 (1%)	194 (0.5%)	<0.0001
	Children in state care	0 (0%)	0 (0%)	1 (0.1%)	0 (0%)	2 (0.1%)	0 (0%)	3 (0.6%)	9 (0.3%)	6 (0.1%)	19 (0.1%)	<0.0001
	Demobilized	0 (0%)	0 (0%)	0 (0%)	0 (0%)	3 (0.2%)	0 (0%)	3 (0.6%)	9 (0.3%)	4 (0.1%)	16 (0%)	<0.0001
	Victims of armed conflict	1 (0.1%)	4 (0.4%)	2 (0.2%)	0 (0%)	5 (0.3%)	6 (1.2%)	4 (0.8%)	9 (0.3%)	20 (0.4%)	33 (0.1%)	<0.0001

Where Bajo Cauca (BC), Magdalena Medio (MM), Nordeste (NE), Norte (NO), Occidente (OC), Oriente (OR), not georeferenced (NG), Suroeste (SO), Urabá (UR), and Valle de Aburrá (VA).

3.2. Symptomatological Behavior by both Subregion and by Type of Dengue

By subregion, all medical and symptomatological variables were significant according to the KW and chi-squared tests. The "medical consultation time" (in days) is when a person starts having symptoms and consults in a medical center. The "clinical deterioration time" (in days) is the time between when a person enters the center and when they are transferred to hospitalization (Table 6).

According to the MW test p-values (Table 7), there was a statistically significant difference between the subregions in terms of medical consultation time, except for the following pairs: Bajo Cauca–Oriente, Suroeste–Magdalena Medio, Nordeste–Occidente, Norte–Urabá, Norte–Valle de Aburrá, and Suroeste–Oriente. Moreover, there were statistically significant differences in clinical deterioration time between the subregions Suroeste–Bajo Cauca, Valle de Aburrá–Occidente, and Magadalena Medio and the rest (except Occidente and Valle de Aburrá–Oriente). Similar symptoms occurred in all subregions.

People suffered from common dengue symptoms, and the proportions by subregion were close to each other. Relevant findings were the rate of hospitalization and severe dengue events, with Urabá and Bajo Cauca having the highest percentages of hospitalized patients, for either normal or severe dengue, with 57.7% and 48.4%, respectively. In contrast, Suroeste and Valle de Aburrá had the smallest percentages of severe dengue cases (0.4%), about half compared to the other subregions (Table 6). Interestingly, Valle de Aburrá is the most populated subregion of Antioquia, within which the city of Medellín is located.

According to the type of dengue (Table 8), there were 50,101 people with dengue (99.4%) and 296 people with severe dengue (0.6%). The percentage of men and women with dengue and severe dengue was the same. The number of adults with dengue was greater than adults with severe dengue (41.3% versus 29.7%, p-value < 0.0001). Furthermore, the number of older people with dengue was less than that of older people with severe dengue (8.2% versus 11.8%, p-value $= 0.03$). There were no statistical differences in the other age group categories in both dengue-type groups.

The medical consultation time of people with dengue was statistically less than people with severe dengue (four days versus three days, p-value < 0.0001), even considering the median bootstrap confidence intervals. Nonetheless, the clinical deterioration time between the two groups was not statistically different. The greatest differences in both groups were in the hospitalization requirement. More people with severe dengue were hospitalized compared to those with normal dengue (98% versus 29.3%, p-value < 0.0001). Regarding symptomatology, fever, retro-ocular pain, myalgia, and arthralgia were not statistically different between the dengue and severe dengue groups. Moreover, headache (86.1% versus 79.7%, p-value $= 0.002$) and rash (48.1% versus 38.5%, p-value $= 0.001$) were statistically different symptoms whose frequency was greater in normal dengue patients than in severe dengue patients (Table 8).

For the remaining symptoms, that is, abdominal pain (26% versus 73.6%), vomiting (22.8% versus 54.4%), diarrhea (15% versus 32.4%), drowsiness (3.2% versus 22%), hypotension (1.5% versus 28%), hepatomegaly (1.2% versus 13.5%), oral ecchymosis (3.6% versus 19.6%), hypothermia (0.4% versus 6.8%), thrombocytopenia (21.5% versus 73%), and high hematocrits levels (3.1% versus 23.6%), there were statistically significant differences between both dengue groups, all these comparisons having a p-value < 0.0001 (Table 8).

It is essential to point out that these last symptoms were more frequent in the severe dengue group than in the normal one, achieving values at least twice larger than in the normal group. Thus, it would be expected that patients presenting these symptoms are likely related to future severe dengue conditions.

Table 6. Medical and symptomatically characteristics among patients, according to subregion location.

Variable		BC (n = 1874)	MM (n = 908)	NE (n = 1218)	NO (n = 475)	OC (n = 1644)	OR (n = 515)	NG (n = 491)	SO (n = 2741)	UR (n = 5196)	VA (n = 35,335)	p-Value
Medical consultation time (in days)		3 (3–4)	3 (3–3)	2 (2–3)	4 (3–4)	2 (2–3)	3 (3–4)	4 (3–4)	3 (3–3)	4 (4–4)	4 (4–4)	<0.0001
Hospitalized patients		907 (48.4%)	362 (39.9%)	457 (37.5%)	191 (40.2%)	426 (25.9%)	218 (42.3%)	99 (20.2%)	660 (24.1%)	3000 (57.7%)	8640 (24.5%)	<0.0001
Severe dengue		20 (1.1%)	11 1.2%)	17 (1.4%)	4 (0.8%)	13 (0.8%)	8 (1.6%)	3 (0.6%)	11 (0.4%)	66 (1.3%)	143 (0.4%)	<0.0001
Clinical deterioration time (in days)		4 (4–4)	4 (3–4)	4 (4–5)	5 (4–5)	4 (4–4)	4 (4–5)	5 (4–5)	4 (4–5)	4 (4–4)	5 (4–5)	<0.0001
Symptoms	Fever	1874 (100%)	908 (100%)	1218 (100%)	475 (100%)	1644 (100%)	515 (100%)	491 (100%)	2741 (100%)	5194 (99.9%)	35328 (99.9%)	<0.0001
	Headache	1651 (88.1%)	815 (89.8%)	964 (79.1%)	412 (86.7%)	1407 (85.6%)	425 (82.5%)	425 (86.6%)	2422 (88.4%)	4702 (90.5%)	30168 (85.4%)	<0.0001
	Retro-ocular pain	794 (42.4%)	482 (53.1%)	519 (42.6%)	241 (50.7%)	762 (46.4%)	239 (46.4%)	315 (64.2%)	1471 (53.7%)	2390 (46%)	17030 (48.2%)	<0.0001
	Myalgia	1558 (83.1%)	745 (82%)	1008 (82.8%)	419 (88.2%)	1430 (87%)	455 (88.3%)	442 (90%)	2384 (87%)	4331 (83.4%)	30595 (86.6%)	<0.0001
	Arthralgia	1354 (72.3%)	663 (73%)	876 (71.9%)	372 (78.3%)	1337 (81.3%)	398 (77.3%)	408 (83.1%)	2182 (79.6%)	3719 (71.6%)	27202 (77%)	<0.0001
	Rash	552 (29.5%)	327 (36%)	455 (37.4%)	201 (42.3%)	754 (45.9%)	246 (47.8%)	314 (64%)	1259 (45.9%)	1604 (30.9%)	18494 (52.3%)	<0.0001
	Abdominal pain	766 (40.9%)	347 (38.2%)	356 (29.2%)	94 (19.8%)	380 (23.1%)	161 (31.3%)	83 (16.9%)	653 (23.8%)	2231 (42.9%)	8166 (23.1%)	<0.0001
	Vomiting	624 (33.3%)	315 (34.7%)	301 (24.7%)	95 (20%)	336 (20.4%)	130 (25.2%)	92 (18.7%)	614 (22.4%)	2066 (39.8%)	7013 (19.8%)	<0.0001
	Diarrhea	256 (13.7%)	155 (17.1%)	165 (13.5%)	56 (11.8%)	202 (12.3%)	97 (18.8%)	57 (11.6%)	388 (14.2%)	1082 (20.8%)	5146 (14.6%)	<0.0001
	Drowsiness	109 (5.8%)	48 (5.3%)	58 (4.8%)	19 (4%)	39 (2.4%)	30 (5.8%)	13 (2.6%)	105 (3.8%)	309 (5.9%)	934 (2.6%)	<0.0001
	Hypotension	54 (2.9%)	18 (2%)	35 (2.9%)	13 (2.7%)	32 (1.9%)	13 (2.5%)	4 (0.8%)	53 (1.9%)	124 (2.4%)	477 (1.3%)	<0.0001
	Hepatomegaly	33 (1.8%)	13 (1.4%)	30 (2.5%)	8 (1.7%)	23 (1.4%)	15 (2.9%)	5 (1%)	54 (2%)	128 (2.5%)	310 (0.9%)	<0.0001
	Oral ecchymosis	87 (4.6%)	28 (3.1%)	47 (3.9%)	16 (3.4%)	67 (4.1%)	26 (5%)	9 (1.8%)	104 (3.8%)	133 (2.6%)	1361 (3.9%)	<0.0001
	Hypothermia	16 (0.9%)	3 (0.3%)	16 (1.3%)	0 (0%)	14 (0.9%)	5 (1%)	2 (0.4%)	19 (0.7%)	20 (0.4%)	137 (0.4%)	<0.0001
	Thrombocytopenia	764 (40.8%)	213 (23.5%)	323 (26.5%)	144 (30.3%)	340 (20.7%)	156 (30.3%)	61 (12.4%)	592 (21.6%)	1880 (36.2%)	6519 (18.4%)	<0.0001
	High hematocrit level	76 (4.1%)	27 (3%)	45 (3.7%)	29 (6.1%)	57 (3.5%)	39 (7.6%)	11 (2.2%)	145 (5.3%)	146 (2.8%)	1065 (3%)	<0.0001

Where Bajo Cauca (BC), Magdalena Medio (MM), Nordeste (NE), Norte (NO), Occidente (OC), Oriente (OR), not georeferenced (NG), Suroeste (SO), Urabá (UR), and Valle de Aburrá (VA).

Table 7. MW test *p*-values for the quantitative variables that rejected the null hypothesis of the KW test by subregions.

Subregion	BC	MM	NE	NO	OC	OR	NG	SO	UR
				Variable "age"					
MM	0.0001	-	-	-	-	-	-	-	-
NE	0.001	1	-	-	-	-	-	-	-
NO	<0.0001	<0.0001	<0.0001	-	-	-	-	-	-
OC	<0.0001	<0.0001	<0.0001	1	-	-	-	-	-
OR	<0.0001	<0.0001	<0.0001	1	0.52	-	-	-	-
NG	<0.0001	<0.0001	<0.0001	1	1	1	-	-	-
SO	<0.0001	<0.0001	<0.0001	1	0.52	0.007	0.47	-	-
UR	<0.0001	<0.0001	<0.0001	<0.0001	<0.0001	<0.0001	<0.0001	<0.0001	-
VA	<0.0001	<0.0001	<0.0001	1	1	0.66	1	0.0003	<0.0001
				Variable "medical consultation time"					
MM	<0.0001	-	-	-	-	-	-	-	-
NE	<0.0001	0.002	-	-	-	-	-	-	-
NO	0.02	<0.0001	<0.0001	-	-	-	-	-	-
OC	<0.0001	0.04	1	<0.0001	-	-	-	-	-
OR	1	0.04	<0.0001	0.04	<0.0001	-	-	-	-
NG	0.11	<0.0001	<0.0001	1	<0.0001	0.11	-	-	-
SO	0.01	0.11	<0.0001	<0.0001	<0.0001	0.97	<0.0001	-	-
UR	<0.0001	<0.0001	<0.0001	1	<0.0001	<0.0001	1	<0.0001	-
VA	<0.0001	<0.0001	<0.0001	1	<0.0001	0.01	1	<0.0001	<0.0001
				Variable "clinical deterioration length time"					
MM	0.67	-	-	-	-	-	-	-	-
NE	0.57	0.01	-	-	-	-	-	-	-
NO	0.08	0.002	1	-	-	-	-	-	-
OC	1	0.29	1	0.87	-	-	-	-	-
OR	1	0.04	1	1	1	-	-	-	-
NG	0.05	0.003	1	1	0.41	1	-	-	-
SO	0.02	0.0001	1	1	1	1	1	-	-
UR	0.05	0.0002	1	1	1	1	0.74	1	-
VA	<0.0001	<0.0001	0.67	1	0.003	1	1	0.74	<0.0001

Where Bajo Cauca (BC), Magdalena Medio (MM), Nordeste (NE), Norte (NO), Occidente (OC), Oriente (OR), not georeferenced (NG), Suroeste (SO), Urabá (UR), and Valle de Aburrá (VA).

3.3. Impact of Socio-Demographic Variables in Clinical Deterioration Time of Hospitalized Patients

A total of 14,960 people were hospitalized (Table 8). We formulate a robust Cox regression to model the impact of some socio-demographic variables on the hazard rate function of the response variable "clinical deterioration time". In this model, the covariates were: sex (male/female), type of dengue (normal/severe), type of settlement (municipal capital/populated center/rural-dispersed), and subregion (all nine).

In a previous descriptive analysis, we found that the behavior of the variable "clinical deterioration time" was common among age groups. In addition, the first and third quartiles as well as the median values were very similar (3, 6, and 4, respectively). Therefore, we decided not to adjust the Cox regression on age. Furthermore, the inclusion of more binary variables in this regression could affect the estimates due to the decrease in the degrees of freedom.

Table 8. Social, medical, and symptomatological characteristics among patients according to dengue type.

	Variable	Dengue (n = 50,101)	Severe Dengue (n = 296)	*p*-Value
	Age	23 (20–25)	26 (26–26)	0.14
Age group	Early childhood (0–5)	3924 (7.8%)	29 (9.8%)	0.2
	Childhood (6–11)	5541 (11.1%)	40 (13.5%)	0.2
	Adolescence (12–18)	7477 (14.9%)	51 (17.2%)	0.3
	Early adulthood (19–26)	8359 (16.7%)	53 (17.9%)	0.6
	Adulthood (27–59)	20,669 (41.3%)	88 (29.7%)	<0.0001
	Old age (60+)	4131 (8.2%)	35 (11.8%)	0.03
Sex	Female	25,190 (50.3%)	149 (50.3%)	1
	Male	24,911 (49.7%)	147 (49.7%)	
Clinical variables	Medical consultation time (in days)	3 (3–4)	4 (4–5)	<0.0001
	Hospitalized patients	14,670 (29.3%)	290 (98%)	<0.0001
	Clinical deterioration time (in days)	4 (4–5)	5 (4-5)	0.27
Symptoms	Fever	50,092 (100%)	296 (100%)	1
	Headache	43,155 (86.1%)	236 (79.7%)	0.002
	Retro-ocular pain	24,099 (48.1%)	144 (48.6%)	0.9
	Myalgia	43,112 (86.1%)	255 (86.1%)	1
	Arthralgia	38,275 (76.4%)	236 (79.7%)	0.2
	Rash	24,092 (48.1%)	114 (38.5%)	0.001
	Abdominal pain	13,019 (26%)	218 (73.6%)	<0.0001
	Vomiting	11,425 (22.8%)	161 (54.4%)	<0.0001
	Diarrhea	7508 (15%)	96 (32.4%)	<0.0001
	Drowsiness	1599 (3.2%)	65 (22%)	<0.0001
	Hypotension	740 (1.5%)	83 (28%)	<0.0001
	Hepatomegaly	579 (1.2%)	40 (13.5%)	<0.0001
	Oral ecchymosis	1820 (3.6%)	58 (19.6%)	<0.0001
	Hypothermia	212 (0.4%)	20 (6.8%)	<0.0001
	Thrombocytopenia	10,776 (21.5%)	216 (73%)	<0.0001
	High hematocrit level	1570 (3.1%)	70 (23.6%)	<0.0001

The baseline levels for the binary variables sex and type of dengue were female and normal, respectively. Regarding the remaining multilevel variables, we structured them in a binary form. In Table 9, we summarize this information, where the target levels are presented in brackets. According to the robust proportional hazard model, only the type of dengue variable was not statistically significant (*p*-value = 0.139) for describing the hazard rate (Table 9). This implies that the type of dengue does not influence the probability that a person will require more days to deteriorate and be admitted to the hospital. The sex variable was significant (*p*-value = 0.013) with an estimated regression coefficient of 0.047, which shows that, considering the other variables as fixed, men have an impact of 1.048 in the estimated hazard rate; that is, male patients are more likely to have a faster clinical deterioration than women. Therefore, women require more days to show clinical deterioration than men (Figure 5a).

Regarding the variable type of settlement, it was statistically significant for the binary relation "populated center–not populated center" (*p*-value = 0.001), with an estimated regression coefficient of 0.120, indicating that, fixing the other variables, people that live in a populated center have an impact of 1.127 in the estimated hazard rate; that is, people in populated centers are more likely to have a faster clinical deterioration compared to people that do not live in a populated center (Figure 5b). Moreover, for the subregion variable, five binary relations were statistically significant (Table 9). The relation "Magdalena Medio–not Magdalena Medio" was the only one with a positive estimated coefficient (0.154).

Thus, people living in a Magdalena Medio have the highest probability of instant clinical deterioration compared to who do not live in that area. The whole comparison between subregions, in terms of the estimated survival functions, is shown in Figure 5b.

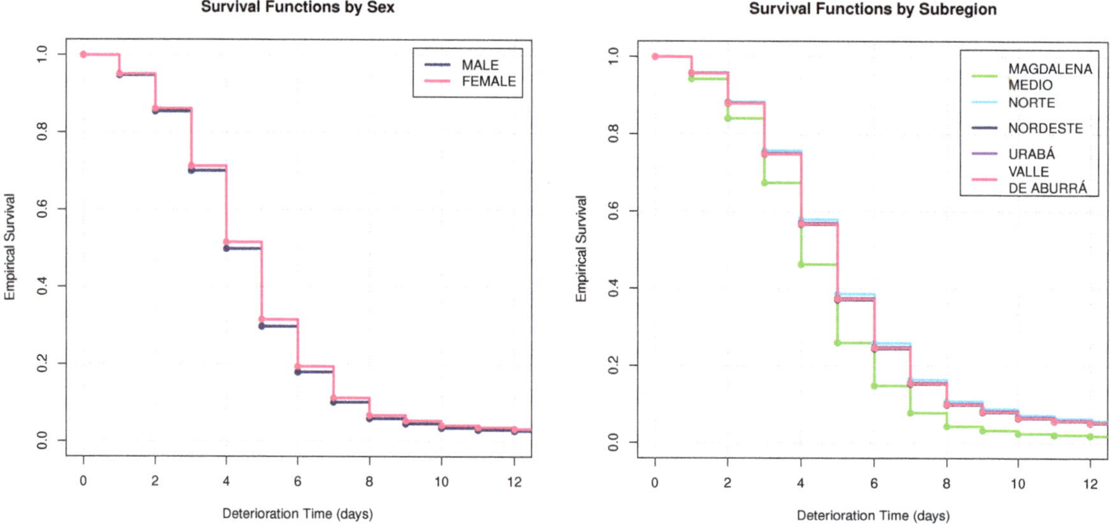

(a) Empirical survival functions by sex.   (b) Empirical survival functions by subregion.

Figure 5. Survival functions for deterioration time according to the robust Cox regression model presented in Table 9.

Table 9. Results of the robust Cox regression for clinical deterioration time analysis using the extended Wald test *p*-value < 0.0001.

Variable	Coefficient	Exp(Coefficient)	SE	*p*-Value
Sex (male)	0.047	1.048	0.019	0.013
Type of dengue (severe)	−0.104	0.902	0.070	0.139
Type of settlement (populated center)	0.120	1.127	0.037	0.001
Type of settlement (rural–dispersed)	−0.010	0.990	0.032	0.760
Subregion (MM)	0.154	1.166	0.073	0.036
Subregion (NE)	−0.153	0.858	0.067	0.022
Subregion (NO)	−0.192	0.826	0.093	0.039
Subregion (OC)	0.013	1.014	0.068	0.843
Subregion (OR)	−0.129	0.879	0.087	0.137
Subregion (SO)	−0.073	0.930	0.059	0.215
Subregion (UR)	−0.164	0.848	0.044	<0.0001
Subregion (VA)	−0.156	0.856	0.041	<0.0001

Where standard error (SE), Magdalena Medio (MM), Nordeste (NE), Norte (NO), Occidente (OC), Oriente (OR), Suroeste (SO), Urabá (UR), and Valle de Aburrá (VA).

4. Discussion

4.1. Dengue in Colombia

Dengue virus is one of the most important arboviral infections worldwide because of its incidence in tropical and subtropical regions [2]. Among these regions, Colombia is one of the most important, as it presents multiple epidemics and hotspot zones. Thus, any social, epidemiological, and medical information about the incidence of the dengue virus in Colombia is crucial in future research for understanding and designing control policies, considering that Colombia has the highest medical cost per day, followed by Vietnam and

Thailand [35]. In this research, we presented a retrospective study of dengue's impact on one of Colombia's most affected zones from 2015 to 2020, the Department of Antioquia.

Regarding how relevant it is to contrast regions systematically, note that, in Colombia, the department of Antioquia is the most critical region in terms of dengue incidence. To the best of our knowledge, detailed analyses of a specific area at the regional and subregional levels have not been carried out so far. Therefore, we consider it an important issue to study this phenomenon at the subregional level due to the potential impact it can generate.

4.2. Socio-Economic State of Dengue in Antioquia

Different factors affect the propagation of the dengue virus over a susceptible population. These are related to health conditions, access to essential services, and socio-economic conditions, which in turn are related to the vector life cycle and proliferation [7]. The assumption of constant mortality and birth rates of mosquitoes is not suitable. This is because they vary depending on environmental conditions [36,37] and variations in both temperature and rain levels, which could increase or decrease these rates [38], as well as affect the bite rate and incubation periods of mosquito offspring [39]. For Antioquia, some subregions that reported outbreaks during 2015–2016 were experiencing one of the most intense ENSO phenomena reported in last years [40], and the localities affected during 2018–2020 were experiencing a moderate ENSO period [41].

Arbovirus transmission mainly comes from urban areas with high population density, medium-low economic class, and poor infrastructure [42] that facilitates *Aedes* breeding sites, for example, water supplies or sewage systems [7]. As pointed out in [43–45], socio-economic factors such as proximity to stagnant waters, poverty, invasions, localized violence, and military migration are some statistically significant risk factors that contribute to a high endemicity. According to our findings, Urabá, Bajo Cauca, and Magdalena Medio had the highest values of dengue prevalence (greater than 50%) over six years (Figure 2). These first two subregions are characterized by having high poverty, overcrowding, and misery rates, as discussed next.

Further information is provided in Table 10 related to urban, rural, and total population indices. This table reports the percentage of people or households that belong to the following specific categories:

(i)   Poverty indicates the percentage of people that cannot pay for essential resources.
(ii) The health barrier shows the percentage of individuals or families that cannot access health services in hospitals.
(iii) No access to water measures the percentage of households with no access to an adequate water supply, such as potable water.
(iv) Overcrowding measures homes with over three people per room, counting the living rooms and dining room but excluding bathrooms, garages, and rooms used for businesses.

In Antioquia, the most affected populations were:

(i)   Adulthood, a working age that represents 44.3% of the total population [20].
(ii) People in elementary occupations (all subregions).
(iii) Displaced, a minority representing 1.1% of the total population in Antioquia; in addition, the term displaced is associated with victims of the Colombian armed conflict, another population representing almost 20% of the population of the whole department [20,21].
(iv) Afro-Colombians in Oriente and Urabá; the last region is this community's major settlement, and 36% of its population lives in rural zones [46].
(v)   Immigrant groups, where 81% of the population is made up of people from Venezuela, followed by people from the United States and Ecuador [46].
(vi) Children in state care (all subregions).

Table 10. Distribution of the population corresponding to the demographic groups in each subregion: 46% and 64% of the urban and rural population in CB are in poverty, respectively. However, because each settlement has a different population size, the percentage of the total does not sum to 100%. Source: Departamento Administrativo Nacional de Estadística (DANE, Spanish acronym), www.dane.gov.co (accessed on 31 January 2022).

Subregion	BC	MM	NE	NO	OC	OR	SO	UR	VA
Poverty									
Urban	46%	28%	27%	24%	24%	17%	24%	40%	10%
Rural	67%	48%	56%	53%	52%	36%	47%	71%	22%
Total	56%	35%	42%	41%	43%	31%	37%	59%	12%
Health barrier									
Urban	4%	4%	2%	3%	3 %	3%	3%	6%	3%
Rural	4%	3%	4%	4%	3 %	2%	3%	5%	3%
Total	4%	3%	4%	4%	3 %	3%	4%	5%	3%
No access to drinking water									
Urban	8%	2%	2%	2 %	1%	1%	1 %	5 %	1%
Rural	36%	23%	60%	60%	29%	37%	41%	70%	18%
Total	19%	12%	32%	26%	16%	14%	21%	43%	3%
Overcrowding									
Urban	19%	10%	8%	7%	9%	6%	6%	15%	4%
Rural	15%	6%	6%	6%	7%	4%	4%	14%	3%
Total	17%	7%	6%	7 %	7%	5%	5%	15%	4%

Where Bajo Cauca (BC), Magdalena Medio (MM), Nordeste (NE), Norte (NO), Occidente (OC), Oriente (OR), Suroeste (SO), Urabá (UR), and Valle de Aburrá (VA).

These communities are vulnerable populations exposed to precarious conditions and forced migrations from other precarious localities. There is a positive relationship between prevalence, land ownership, migration, and forced displacement. This is especially detected in Urabá and Oriente, the most important agricultural production subregions [47], which have been strongly affected by the events of the Colombian armed conflict [48].

Regarding dengue vaccination, currently, there are some licensed vaccines, for example, CYD-TDV, Dengvaxia, Sanofi Pasteur, and candidates such as TAK-003 [49,50]. The last one has been clinically tried in Colombia and presents high efficiency for DENV-2, but not for other serotypes [49]. The idea of developing immunization campaigns for vulnerable populations is closer to being realized in Colombia. Even so, it is important to point out that Colombia is a hyperendemic country in which the predominant serotype may change [39]. Thus, the vaccination process must consider serotype co-circulation analyses to guarantee proficient immunization.

4.3. Socio-Demographic Hazards and Relationship of Dengue and Severe Dengue Symptoms

At a general level in Antioquia, we identified no significant difference between men and women in the prevalence of dengue and severe dengue (Table 8). However, the sex prevalence changes at the subregional level. For example, in Bajo Cauca, Magdalena Medio, and Nordeste, a high male prevalence was observed; see Tables 2 and 4 for the sex ratio per subregion. In [51], it was reported that severe dengue is mostly female-biased after puberty and unbiased for the rest of the age classes. Thus, there is no unique pattern for other localities in the world. For example, Singapore reported no sex bias [52], while Pakistan [53] has a high prevalence for men, and Brazil [54] and Nicaragua [55] have higher prevalences for females. We detected that working ages in Antioquia (adults between 27 and 86 years old) are the most affected group, similar to the results reported in some localities in Brazil [54], Pakistan [56], and Saudi Arabia [57], where the average age of affected adults ranges between 22 and 52 years [54]. We could explain this trend from the socio-economical approach: working ages are high-mobility groups that travel outside their neighborhood, spending over seven hours per day in educational institutions or working areas. This dynamic increases the probability of dengue, especially in high-mosquito-density areas [57].

We assessed that hemorrhagic manifestations are also adult-biased in Antioquia, where approximately 60% of the severe cases are adults above 18 years, as reported in [53,54]. Primary infection is a protective factor against the hemorrhagic forms in children since repeated infections by a different serotype of the dengue result in more severe manifestations [51,54]. The most common symptoms in Antioquia are almost the same: fever, headache, retro-ocular pain, myalgia, arthralgia, and rash, as reported in [53,54]. The hospitalization rate in Antioquia is 29.6% of the cases over the last five years, with Bajo Cauca and Urabá being the most affected subregions. Other studies reported a lower hospitalization, for example, 13.2% of the reported cases [54].

The main results of survival analysis with the robust Cox regression showed that the factors that affect the probability of instant clinical deterioration are sex and location. According to our findings, men suffer from a faster clinical deterioration than women. In addition, people that live in crowded areas, such as Magdalena Medio, mainly characterized by high poverty levels, bad sanitary conditions, informal employment (elementary occupations in general), and vulnerable populations (displaced, immigrants, or children in state care), suffer a faster clinical deterioration. This is an important result on the topic as survival analyses in dengue have been principally implemented for other studies, such as the rate of healing on severe dengue patients [58,59] and survival rates [56].

4.4. Dengue Infections with the COVID-19 Pandemic

The year 2020 was remarkable for the study of epidemiology. The control of COVID-19's propagation affected the incidence, treatment, and identification of other infectious diseases.

After the effects of COVID-19 on the total population, the World Health Organization reported over 1.6 million arboviral cases in the Americas, with the majority (about 97%) being dengue. At the end of 2019 and the beginning of 2020, a dengue outbreak started in Colombia [11]. Its prevalence was less than that reported in the same period in 2019 [60,61]. Some subregions of Antioquia, such as Magdalena, Nordeste, Bajo Cauca, Urabá, and Occidente, presented a decrease in incidence; see Table 3 and Figure 3. Some recent studies relate the decrease in dengue cases to the effect of COVID-19 on health facilities and the diagnosis becoming problematic because both diseases exhibit similar clinical and laboratory manifestations [62]. Other authors attribute the decrease in dengue cases to the lockdowns and declines in regional migration [61]. In [63], a strong association was highlighted between COVID-19-related social changes and the reduction in dengue transmission (school closures and reduced time spent in nonresidential areas). This finding shows evidence that dengue transmission occurred in shared areas outside the home [61].

5. Conclusions

Dengue virus is transmitted by *Aedes* species and presents hyperendemic behaviors in tropical-subtropical regions. Colombia is one of the most affected countries in the Americas. The central-west region is a hot spot in dengue transmission, especially the subtropical localities of the Department of Antioquia. This zone has suffered multiple dengue outbreaks recently (2015–2016 and 2019–2020). As dengue is a disease of high interest to public health in the affected localities, we have formulated Cox regression models and conducted statistical analyses to identify the hazard and socio-demographic patterns of this infectious disease in a Colombian subtropical region between the years of 2015 and 2020. Hence, we performed a retrospective analysis of the confirmed dengue cases in Antioquia by discriminating by both subregions and dengue severity during these years. First, we conducted an exploratory analysis of the epidemic data, and then we formulated a statistical survival analysis using a Cox regression model. Our findings allowed the identification of the hazard and socio-demographic patterns of dengue infections in Antioquia, Colombia, from 2015 to 2020.

We studied the clinical deterioration time. A possible future work might be related to a survival analysis for the time to medical consultation. A comparison of our results with other similar works in the South American region, including Andean countries, is proposed for future research. This will allow us to have a global vision of Dengue and not a local country vision, given the importance of this disease in the Americas as endemic.

Author Contributions: Data curation S.O., A.C.-L., H.V., J.P.R. and A.P.-C. Conceptualization, A.C.-L., H.V., J.P.R., A.P.-C., H.L. and V.L. Formal analysis, S.O., A.C.-L., H.V., J.P.R., A.P.-C., H.L. and V.L. Investigation, S.O., A.C.-L., H.V., J.P.R., A.P.-C., H.L. and V.L. Methodology, S.O., A.C.-L., H.V., J.P.R., A.P.-C., H.L. and V.L. Writing—original draft, S.O., A.C.-L., H.V., J.P.R., A.P.-C. and H.L. Writing—review and editing, V.L. All authors have read and agreed to the published version of the manuscript.

Funding: This research was partially funded by Universidad EAFIT, grant number 954-000002 (S.O., A.C.-L., H.L.) and by Ministerio de Ciencia Tecnología e Innovación de Colombia, projects (I) Formación de Capital Humano de Alto Nivel–Universidad EAFIT—Corte 2 Nacional, BPIN 2020000100778 (H.V.), (II) Convocatoria 757 2016 Doctorados Nacionales FP44842-061-2018 (A.P.-C.) and (III) Convocatoria 909-2 2022 (S.O., A.C.-L.). This research was also partially funded by project grant FONDECYT 1200525 (V.L.) from the National Agency for Research and Development (ANID) of the Chilean government under the Ministry of Science, Technology, Knowledge, and Innovation. The funders had no role in study design, data collection and analysis, publication decisions, or manuscript preparation.

Institutional Review Board Statement: Not applicable.

Informed Consent Statement: Not applicable.

Data Availability Statement: Not applicable.

Acknowledgments: The authors would also like to thank five reviewers for their constructive comments which led to improve the presentation of the manuscript. The authors also thank the Secretaria Seccional de Salud y Protección Social de Antioquia, Colombia, for providing information and the datasets.

Conflicts of Interest: There are no conflict of interest declared by the authors.

References

1. Calderón, A.; Guzmán, C.; Oviedo-Socarras, T.; Mattar, S.; Rodríguez, V.; Castañeda, V.; Moraes Figueiredo, L.T. Two cases of natural infection of dengue-2 virus in bats in the Colombian Caribbean. *Trop. Med. Infect. Dis.* **2021**, *6*, 35. [CrossRef] [PubMed]
2. Ioos, S.; Mallet, H.P.; Goffart, I.L.; Gauthier, V.; Cardoso, T.; Herida, M. Current zika virus epidemiology and recent epidemics. *Med. Mal. Infect.* **2014**, *44*, 302–307. [CrossRef]
3. Mordecai, E.A.; Cohen, J.M.; Evans, M.V.; Gudapati, P.; Johnson, L.R.; Lippi, C.A.; Miazgowicz, K.; Murdock, CC; Rohr, J.R.; Ryan, S.J.; et al. Detecting the impact of temperature on transmission of zika, dengue, and chikungunya using mechanistic models. *PLoS Negl. Trop. Dis.* **2017**, *11*, e0005568. [CrossRef] [PubMed]
4. Wang, H.; Naghavi, M.; Allen, C.; Barber, R.M.; Bhutta, Z.A.; Carter, A.; Casey, D.C.; Charlson, F.J.; Chen, A.Z.; Coates, M.M.; et al. Global, regional, and national life expectancy, all-cause mortality, and cause-specific mortality for 249 causes of death, 1980–2015: A systematic analysis for the global burden of disease study 2015. *Lancet* **2016**, *388*, 1459–1544. [CrossRef] [PubMed]
5. Desjardins, M.R.; Casas, I.; Victoria, A.M.; Carbonell, D.; Dávalos, D.M.; Delmelle, E.M. Knowledge, attitudes, and practices regarding dengue, chikungunya, and zika in Cali, Colombia. *Health Place* **2020**, *63*, 102339. [CrossRef]
6. Mora-Salamanca, A.F.; Porras-Ramírez, A.; De la Hoz Restrepo, F.P. Estimating the burden of arboviral diseases in Colombia between 2013 and 2016. *Int. J. Infect. Dis.* **2020**, *97*, 81–89. [CrossRef]
7. Villar, L.A.; Rojas, D.P.; Besada-Lombana, S.; Sarti, E. Epidemiological trends of dengue disease in Colombia (2000–2011): A systematic review. *PLoS Negl. Trop. Dis.* **2015**, *9*, e0003499. [CrossRef] [PubMed]
8. Padilla, J.C.; Rojas, D.P.; Gómez, R.S. *Dengue en Colombia: Epidemiologia de la Reemergencia y la Hiperendemia*; Ministerio de la Protección Social-Instituto Nacional de Salud: Bogotá, Colombia, 2012.
9. Carabali, M.; Jaramillo-Ramirez, G.I.; Rivera, V.A.; Mina Possu, N.J.; Restrepo, B.N.; Zinszer, K. Assessing the reporting of dengue, chikungunya and zika to the national surveillance system in Colombia from 2014–2017: A capture-recapture analysis accounting for misclassification of arboviral diagnostics. *PLoS Neglected Trop. Dis.* **2021**, *15*, e0009014. [CrossRef] [PubMed]

10. Rico-Mendoza, A.; Alexandra, P.R.; Chang, A.; Encinales, L.; Lynch, R. Co-circulation of dengue, chikungunya, and zika viruses in colombia from 2008 to 2018. *Rev. Panam. Salud Pública* **2019**, *43*, 1. [CrossRef]
11. Cardona-Ospina, J.A.; Arteaga-Livias, K.; Villamil-Gómez, W.E.; Pérez-Díaz, C.E.; Katterine Bonilla-Aldana, D.; Mondragon-Cardona, á.; Solarte-Portilla, M.; Martinez, E.; Millan-Oñate, J.; López-Medina, E.; et al. Dengue and COVID-19, overlapping epidemics? An analysis from Colombia. *J. Med Virol.* **2020**, *93*, 522–527. [CrossRef]
12. Aliota, M.T.; Peinado, S.A.; Velez, I.D.; Osorio, J.E. The wMel strain of Wolbachia reduces transmission of zika virus by Aedes Aegypti. *Sci. Rep.* **2016**, *6*, 28792. [CrossRef] [PubMed]
13. Ocampo, C.B.; Salazar-Terreros, M.J.; Mina, N.J.; McAllister, J.; Brogdon, W. Insecticide resistance status of Aedes Aegypti in 10 localities in Colombia. *Acta Trop.* **2011**, *118*, 37–44. [CrossRef] [PubMed]
14. Lima, E.C.B.D.; Montarroyos, U.R.; Magalhães, J.J.F.D.; Dimech, G.S.; Lacerda, H.R. Survival analysis in non-congenital neurological disorders related to dengue, chikungunya and zika virus infections in northeast Brazil. *Rev. Inst. Med. Trop. S. Paulo* **2020**, *62*. [CrossRef]
15. Oviedo-Pastrana, M.; Méndez, N.; Mattar, S.; Arrieta, G.; Gomezcaceres, L. Epidemic outbreak of chikungunya in two neighboring towns in the Colombian caribbean: A survival analysis. *Arch. Public Health* **2017**, *75*, 1. [CrossRef]
16. Pinzón, M.A.; Ortiz, S.; Holguín, H.; Betancur, J.F.; Cardona Arango, D.; Laniado, H.; Arias Arias, C.; Muñoz, B.; Quiceno, J.; Jaramillo, D.; et al. Dexamethasone vs methylprednisolone high dose for COVID-19 pneumonia. *PLoS ONE* **2021**, *16*, e0252057. [CrossRef]
17. Departamento Administrativo de Planeación. Temperatura Promedio Anual, en los Municipios de ANTIOQUIA, 2016. Available online: http://www.antioquiadatos.gov.co/index.php/20-4-1-temperatura-promedio-anual-en-los-municipios-de-antioquia-ano-2015 (accessed on 9 May 2021).
18. Departamento Administrativo de Planeación, Situación Geográfica, Extensión km^2, Altura y Temperatura de los Municipios de Antioquia, por Subregión, 2016. Available online: http://www.antioquiadatos.gov.co/index.php/1-4-1-situacion-geografica-extension-km-altura-y-temperatura-de-los-municipios-de-antioquia-por-subregion (accessed on 9 May 2021).
19. United Nations Development Programme. Antioquia: Retos y desafíos para el Desarrollo Sostenible. 2019. Available online: https://www.co.undp.org/content/colombia/es/home/library/democratic_governance/antioquia--retos-y-desafios-para-el-desarrollo-sostenible.html (accessed on 9 May 2021).
20. Secretaria Seccional de Salud y Protección Social de Antioquia. *Análisis de Situación de Salud Actualización 2020*; Gobernación de Antioquia: Medellín, Colombia, 2020; pp. 1–372.
21. Secretaria Seccional de Salud y Protección Social de Antioquia. *Plataforma Entornos Familiar y Saludable*; Gobernación de Antioquia:Medellín, Colombia, 2021.
22. Secretaria Seccional de Salud y Protección Social de Antioquia. *Política Pública de Discapacidad e Inclusión Social Departamento de Antioquia*; Gobernación de Antioquia: Medellín, Colombia, 2015; pp. 1–94.
23. Ministerio de Salud de Colombia. *Boletines Poblacionales: Personas con Discapacidad*; Gobernación de Antioquia: Medellín, Colombia, 2020; pp. 1–15.
24. Departamento Administrativo Nacional de Estadística. Estadísticas de las Personas Desmovilizadas que han Ingresado al Proceso de Reintegración; Datos Abiertos Colombia, 2022. Available at www.datos.gov.co (accessed on 5 December 2022).
25. Reinhold, J.M.; Lazzari, C.R.; Lahondère, C. Effects of the environmental temperature on Aedes Aegypti and Sedes albopictus mosquitoes: A review. *Insects* **2018**, *9*, 158. [CrossRef] [PubMed]
26. Méndez, J.A.; Usme-Ciro, J.A.; Domingo, C.; Rey, G.J.; Sánchez, J.A.; Tenorio, A.; Gallego-Gomez, J.C. Phylogenetic reconstruction of dengue virus type 2 in Colombia. *Virol. J.* **2012**, *9*, 64. [CrossRef]
27. Quintero-Herrera, L.L.; Ramírez-Jaramillo, V.; Bernal-Gutiérrez, S.; Cárdenas-Giraldo, E.V.; Guerrero-Matituy, E.A.; Molina-Delgado, A.H.; Montoya-Arias, C.P.; Rico-Gallego, J.A.; Herrera-Giraldo, A.C.; Botero-Franco, S.; et al. Potential impact of climatic variability on the epidemiology of dengue in Risaralda, Colombia, 2010–2011. *J. Infect. Public Health* **2015**, *8*, 291–297. [CrossRef]
28. Vincenti-Gonzalez, M.F.; Tami, A.; Lizarazo, E.F.; Grillet, M.E. ENSO-driven climate variability promotes periodic major outbreaks of dengue in Venezuela. *Sci. Rep.* **2018**, *8*, 5727. [CrossRef]
29. Giedion, U.; Uribe, M.V. Colombia's universal health insurance system. *Health Aff.* **2009**, *28*, 853–863. [CrossRef]
30. International Labour Office (ILO). *International Standard Classification of Occupations 2008 (ISCO-08): Structure, Group Definitions and Correspondence Tables*; International Labour Office, Geneva, Switzerland, 2012; Volume 1.
31. R Core Team. *R: A Language and Environment for Statistical Computing*; R Foundation for Statistical Computing: Vienna, Austria, 2022.
32. Minder, C.E.; Bednarski, T. A robust method for proportional hazards regression. *Stat. Med.* **1996**, *15*, 1033–1047. [CrossRef]
33. Cox, D.R. Regression models and life-tables. *J. R. Stat. Soc. B* **1972**, *34*, 187–220. [CrossRef]
34. Bednarski, T.; Borowicz, F. Coxrobust: Robust Estimation in Cox Model; R Package Version 1.0; 2006. Available at cran.r-project.org/web/packages/coxrobust (accessed on 1 November 2022)
35. Lee, J.S.; Mogasale, V.; Lim, J.K.; Carabali, M.; Lee, K.S.; Sirivichayakul, C.; Dang, D.A.; Palencia-Florez, D.C.; Nguyen, T.H.A.; Riewpaiboon, A.; et al. A multi-country study of the economic burden of dengue fever: Vietnam, Thailand, and Colombia. *PLoS Negl. Trop. Dis.* **2017**, *11*, e0006037. [CrossRef] [PubMed]
36. Catano-Lopez, A.; Rojas-Diaz, D.; Laniado, H.; Arboleda-Sánchez, S.; Puerta-Yepes, M.E.; Lizarralde-Bejarano, D.P. An alternative model to explain the vectorial capacity using as example Aedes Aegypti case in dengue transmission. *Heliyon* **2019**, *5*, e02577. [CrossRef]

37. Velasco, H.; Laniado, H.; Toro, M.; Catano-López, A.; Leiva, V.; Lio, Y. Modeling the risk of infectious diseases transmitted by Sedes Aegypti using survival and aging statistical analysis with a case study in Colombia. *Mathematics* **2021**, *9*, 1488. [CrossRef]
38. Jácome, G.; Vilela, P.; Yoo, C. Social-ecological modelling of the spatial distribution of dengue fever and its temporal dynamics in Guayaquil, Ecuador for climate change adaption. *Ecol. Inform.* **2019**, *49*, 1–12. [CrossRef]
39. Gutierrez-Barbosa, H.; Medina-Moreno, S.; Zapata, J.C.; Chua, J.V. Dengue infections in Colombia: Epidemiological trends of a hyperendemic country. *Trop. Med. Infect. Dis.* **2020**, *5*, 156. [CrossRef]
40. Santoso, A.; Mcphaden, M.J.; Cai, W. The defining characteristics of ENSO extremes and the strong 2015/2016 El Niño. *Rev. Geophys.* **2017**, *55*, 1079–1129. [CrossRef]
41. Giraldo-Osorio, J.D.; Trujillo-Osorio, D.E.; Baez-Villanueva, O.M. Analysis of ENSO-driven variability, and long-term changes, of extreme precipitation indices in Colombia, using the satellite rainfall estimates chirps. *Water* **2022**, *14*, 1733. [CrossRef]
42. Fuentes-Vallejo, M. Space and space-time distributions of dengue in a hyper-endemic urban space: The case of Girardot, Colombia. *BMC Infect. Dis.* **2017**, *17*, 512. [CrossRef]
43. Murray, N.E.A.; Quam, M.B.; Wilder-Smith, A. Epidemiology of dengue: Past, present and future prospects. *Clin. Epidemiol.* **2013**, *5*, 299. [PubMed]
44. Krystosik, A.R.; Curtis, A.; Buritica, P.; Ajayakumar, J.; Squires, R.; Dávalos, D.; Pacheco, R.; Bhatta, M.P.; James, M.A. Community context and sub-neighborhood scale detail to explain dengue, chikungunya and zika patterns in Cali, Colombia. *PLoS ONE* **2017**, *12*, e0181208. [CrossRef]
45. Lippi, C.A.; Stewart-Ibarra, A.M.; Muñoz, á.G.; Borbor-Cordova, M.J.; Mejía, R.; Rivero, K.; Castillo, K.; Cárdenas, W.B.; Ryan, S.J. The social and spatial ecology of dengue presence and burden during an outbreak in Guayaquil, Ecuador, 2012. *Int. J. Environ. Res. Public Health* **2018**, *15*, 827. [CrossRef] [PubMed]
46. *Departamento Administrativo Nacional de Estadística*; Censo Nacional de Población y Vivienda: Bogotá, Colombia, 2018.
47. Gaviria, C.F.; Muñoz, J.C. Desplazamiento forzado y propiedad de la tierra en Antioquia, 1996–2004. *Lect. Econ.* **2009**, *66*, 9–46. [CrossRef]
48. Jurisdicción Especial para la Paz (JEP). Macrocaso 04: Situación Territorial de la Región de Urabá. 2022. Available online: https://www.jep.gov.co/especiales1/macrocasos/04.html (accessed on 7 September 2022).
49. López-Medina, E.; Biswal, S.; Saez-Llorens, X.; Borja-Tabora, C.; Bravo, L.; Sirivichayakul, C.; Vargas, L.M.; Alera, M.T.; Velásquez, H.; Reynales, H.; et al. Efficacy of a Dengue Vaccine Candidate (TAK-003) in Healthy Children and Adolescents 2 Years after Vaccination. *J. Infect. Dis.* **2022**, *9*, 1521–1532. [CrossRef]
50. Vaccine, D. WHO Position Paper. *Wkly. Epidemiol. Rec.* **2018**, *93*, 457–476.
51. Guerra-Silveira, F.; Abad-Franch, F. Sex bias in infectious disease epidemiology: Patterns and processes. *PLoS ONE* **2013**, *8*, e62390. [CrossRef]
52. Yung, C.F.; Lee, K.S.; Thein, T.L.; Tan, L.K.; Gan, V.C.; Wong, J.G.; Lye, D.C.; Ng, L.C.; Leo, Y.S. Dengue serotype-specific differences in clinical manifestation, laboratory parameters and risk of severe disease in adults, Singapore. *Am. J. Trop. Med. Hyg.* **2015**, *92*, 999–1005. [CrossRef]
53. Arshad, H.; Bashir, M.; Mushtaq, U.S.; Imtiaz, H.; Rajpar, R.; Alam, M.F.; Fatima, S.; Rehman, A.; Abbas, K.; Talpur, A.S. Clinical characteristics and symptomatology associated with dengue fever. *Cureus* **2022**, *14*, e26677. [CrossRef]
54. Souza, L.J.D.; Pessanha, L.B.; Mansur, L.C.; Souza, L.A.D.; Ribeiro, M.B.T.; Silveira, M.D.V.D.; Souto Filho, J.T.D. Comparison of clinical and laboratory characteristics between children and adults with dengue. *Braz. J. Infect. Dis.* **2013**, *17*, 27–31. [CrossRef]
55. Hammond, S.N.; Balmaseda, A.; Perez, L.; Tellez, Y.; Saborío, S.I.; Mercado, J.C.; Videa, E.; Rodriguez, Y.; Perez, M.A.; Cuadra, R.; et al. Differences in dengue severity in infants, children, and adults in a 3-year hospital-based study in Nicaragua. *Am. J. Trop. Med. Hyg.* **2005**, *73* 1063–1070. [CrossRef] [PubMed]
56. Chaudhry, K.A.; Jamil, F.; Razzaq, M.; Jilani, B.F. Survival analysis of dengue patients of Pakistan. *Int. J. Mosq. Res.* **2018**, *5*, 5–9.
57. Khormi, H.M.; Kumar, L. Modeling dengue fever risk based on socioeconomic parameters, nationality and age groups: GIS and remote sensing based case study. *Sci. Total Environ.* **2011**, *409*, 4713–4719. [CrossRef] [PubMed]
58. Annas, S.; Nusrang, M.; Arisandi, R.; Fadillah, N.; Kartikasari, P. Cox proportional hazard regression analysis of dengue hemorrhagic fever. *J. Phys. Conf. Ser.* **2018**, *1028*, 012242.
59. Irfan, M.; Usman, M.; Saidi, S.; Kurniasari, D. Survival analysis using cox proportional hazard regression approach in dengue hemorrhagic fever (DHF) case in Abdul Moeloek hospital Bandar Lampung in 2019. *J. Phys. Conf. Ser.* **2021**, *1751*, 012011. [CrossRef]
60. Microbe, T.L. Arboviruses and COVID-19: The need for a holistic view. *Lancet Microbe* **2020**, *1*, e136. [CrossRef]
61. Sasmono, R.T.; Santoso, M.S. Movement dynamics: Reduced dengue cases during the COVID-19 pandemic. *Lancet Infect. Dis.* **2022**, *22*, 570–571. [CrossRef]

62. Awan, U.A.; Zahoor, S.; Ayub, A.; Ahmed, H.; Aftab, N.; Afzal, M.S. COVID-19 and arboviral diseases: Another challenge for Pakistan's dilapidated healthcare system. *J. Med Virol.* **2020**, *93*, 4065–4067. [CrossRef]
63. Chen, Y.; Li, N.; Lourenço, J.; Wang, L.; Cazelles, B.; Dong, L.; Li, B.; Liu, Y.; Jit, M.; Bosse, N.I.; et al. Measuring the effects of covid-19-related disruption on dengue transmission in southeast Asia and latin America: A statistical modelling study. *Lancet Infect. Dis.* **2022**, *22*, 657–667. [CrossRef]

Tropical Medicine and Infectious Disease

Article

Co-Radiation of *Leptospira* and Tenrecidae (Afrotheria) on Madagascar

Yann Gomard [1,†], Steven M. Goodman [2,3], Voahangy Soarimalala [2], Magali Turpin [1], Guenaëlle Lenclume [1], Marion Ah-Vane [1], Christopher D. Golden [4,5] and Pablo Tortosa [1,*]

[1] Unité Mixte de Recherche Processus Infectieux en Milieu Insulaire Tropical (UMR PIMIT), Université de La Réunion, CNRS 9192, INSERM 1187, IRD 249, Plateforme Technologique CYROI, 97490 Sainte-Clotilde, France
[2] Association Vahatra, BP 3972, Antananarivo 101, Madagascar
[3] Field Museum of Natural History, Chicago, IL 60605, USA
[4] Department of Nutrition, Harvard TH Chan School of Public Health, Boston, MA 02115, USA
[5] Department of Environmental Health, Harvard TH Chan School of Public Health, Boston, MA 02115, USA
*   Correspondence: pablo.tortosa@univ-reunion.fr
†   Current address: Unité Mixte de Recherche Peuplements Végétaux et Bioagresseurs en Milieu Tropical (UMR PVBMT), Université de La Réunion, Pôle de Protection des Plantes, 97410 Saint-Pierre, France.

Citation: Gomard, Y.; Goodman, S.M.; Soarimalala, V.; Turpin, M.; Lenclume, G.; Ah-Vane, M.; Golden, C.D.; Tortosa, P. Co-Radiation of *Leptospira* and Tenrecidae (Afrotheria) on Madagascar. *Trop. Med. Infect. Dis.* **2022**, *7*, 193. https://doi.org/10.3390/tropicalmed7080193

Academic Editors: Vyacheslav Yurchenko and Thomas Dorlo

Received: 20 June 2022
Accepted: 16 August 2022
Published: 18 August 2022

Publisher's Note: MDPI stays neutral with regard to jurisdictional claims in published maps and institutional affiliations.

Abstract: Leptospirosis is a bacterial zoonosis caused by pathogenic *Leptospira* that are maintained in the kidney lumen of infected animals acting as reservoirs and contaminating the environment via infected urine. The investigation of leptospirosis through a *One Health* framework has been stimulated by notable genetic diversity of pathogenic *Leptospira* combined with a high infection prevalence in certain animal reservoirs. Studies of Madagascar's native mammal fauna have revealed a diversity of *Leptospira* with high levels of host-specificity. Native rodents, tenrecids, and bats shelter several distinct lineages and species of *Leptospira*, some of which have also been detected in acute human cases. Specifically, *L. mayottensis*, first discovered in humans on Mayotte, an island neighboring Madagascar, was subsequently identified in a few species of tenrecids on the latter island, which comprise an endemic family of small mammals. Distinct *L. mayottensis* lineages were identified in shrew tenrecs (*Microgale cowani* and *Nesogale dobsoni*) on Madagascar, and later in an introduced population of spiny tenrecs (*Tenrec ecaudatus*) on Mayotte. These findings suggest that *L. mayottensis* (i) has co-radiated with tenrecids on Madagascar, and (ii) has recently emerged in human populations on Mayotte following the introduction of *T. ecaudatus* from Madagascar. Hitherto, *L. mayottensis* has not been detected in spiny tenrecs on Madagascar. In the present study, we broaden the investigation of Malagasy tenrecids and test the emergence of *L. mayottensis* in humans as a result of the introduction of *T. ecaudatus* on Mayotte. We screened by PCR 55 tenrecid samples from Madagascar, including kidney tissues from 24 individual *T. ecaudatus*. We describe the presence of *L. mayottensis* in Malagasy *T. ecaudatus* in agreement with the aforementioned hypothesis, as well as in *M. thomasi*, a tenrecid species that has not been explored thus far for *Leptospira* carriage.

Keywords: microbial endemism; *Leptospira mayottensis*; leptospirosis; tenrecids; Madagascar; Mayotte

1. Introduction

Leptospirosis is a zoonotic disease that results annually in around 1 million human cases and nearly 60,000 deaths [1]. *Leptospira* bacteria, the pathogen responsible for the disease, are maintained in the lumen of the kidney tubules of animal reservoirs [2], which can chronically shed viable bacteria in their urine and contaminate the environment [3]. Although humans can be affected through direct contact with infected reservoirs, indirect transmission during outdoor activities in a contaminated environment is most frequent [4]. Infection leads to a wide range of symptoms ranging from mild flu-like syndromes to multi-organ failure causing death in 5–10% of the cases.

The genus *Leptospira* is currently composed of more than 60 taxa including saprophytic and pathogenic species [3,5–8]. Investigations carried out in different areas of the world through a *One Health* approach have shown distinct transmission chains composed of species or lineages and reservoirs that vary from one environmental setting to another [9–13]. Investigations carried out in the ecosystems of Madagascar and surrounding islands, hereafter referred to as the Malagasy Region, have provided new information on transmission chains on the different islands [14]. Indeed, on La Réunion and in the Seychelles, human leptospirosis is mostly caused by *Leptospira* that are broadly distributed and hence likely of introduced origin [11,12]. By contrast, Madagascar and Mayotte, a French-administrated island in the Comoros archipelago, shelter distinctly more diversified *Leptospira* assemblages, including species and lineages that are best considered endemic [15–17].

Among pathogenic *Leptospira* described and investigated in the Malagasy Region, *L. mayottensis*, the principal focus of the current study, warrants further characterization. These bacteria were first isolated from acute human leptospirosis cases on Mayotte and initially named *L. borgpetersenii* group B [9,18]. A thorough characterization of serological and genomic features of these isolates led the French Reference Centre on Spirochetes to elevate this bacterium to the rank of a new species, which was named *L. mayottensis* in reference to the geographic origin of the human isolates [19]. A comprehensive investigation of the Malagasy wild mammal fauna allowed identification of *Leptospira* samples imbedded in the genetic clade of *L. mayottensis* and shed by two endemic small mammal species, namely *Microgale cowani* and *Nesogale dobsoni* [20]. These two host species belong to the endemic family Tenrecidae, composed of omnivorous small mammals known to play an important role in *Leptospira* maintenance as reservoirs of two distinct species: *L. borgpetersenii* and *L. mayottensis* [17,20,21]. The origin of the Tenrecidae, a monophyletic group, is the result of a single colonization event originating from Africa that took place 30–56 million years ago, followed by an extraordinary radiation leading to the currently named nearly 40 extant species or confirmed candidate species [22,23]. These findings strongly suggest that *L. mayottensis* has co-radiated with tenrecid hosts on Madagascar.

It has been proposed that *L. mayottensis* was introduced to Mayotte from Madagascar [24]. This was supported by an investigation of animal reservoirs on Mayotte identifying *Tenrec ecaudatus*, a spiny Tenrec introduced from Madagascar for human consumption, as the local reservoir of *L. mayottensis*. However, the hypothesis that *T. ecaudatus* sheds *L. mayottensis* currently lacks definitive evidence for Malagasy populations of this species. In the present investigation, we screened *T. ecaudatus* specimens together with other tenrecid species sampled on Madagascar to broaden information on the presence of *L. mayottensis* in these animals, and to test the hypothesis of *L. mayottensis* being transported to Mayotte associated with the introduction of *T. ecaudatus*.

2. Materials and Methods

2.1. Biological Sample

All investigated shrew tenrecs (subfamily Oryzorictinae) were sampled in February 2016 in a forest neighboring the village of Anjozorobe, in the Central Highlands of Madagascar (see Figure 1). The samples included 31 specimens belonging to the following nine species: *Microgale taiva* (*n* = 15), *M. thomasi* (*n* = 3), *M. majori* (*n* = 3), *M. parvula* (*n* = 2), *M. soricoides* (*n* = 2), *M. cowani* (*n* = 1), *M. longicaudata* (*n* = 1), *M. fotsifotsy* (*n* = 1), and *Nesogale dobsoni* (*n* = 3). The spiny tenrec samples composed of *Tenrec ecaudatus* (subfamily Tenrecinae) included 24 specimens collected in villages adjacent to the Makira Natural Park in the Commune Antsirabe-Sahatany (Maroantsetra District) (Figure 1), an area with heavy human hunting pressure [25]. All samples in this region were collected from captured animals provided by local hunters to the research team. All specimens were captured, manipulated, and euthanized following guidelines accepted by the scientific community for the handling of wild mammals [26] and in strict accordance with permits issued by the national authorities of Madagascar. All kidney samples from the collected animals from both project areas were immediately stored in 70% ethanol until DNA extraction and molecular analyses.

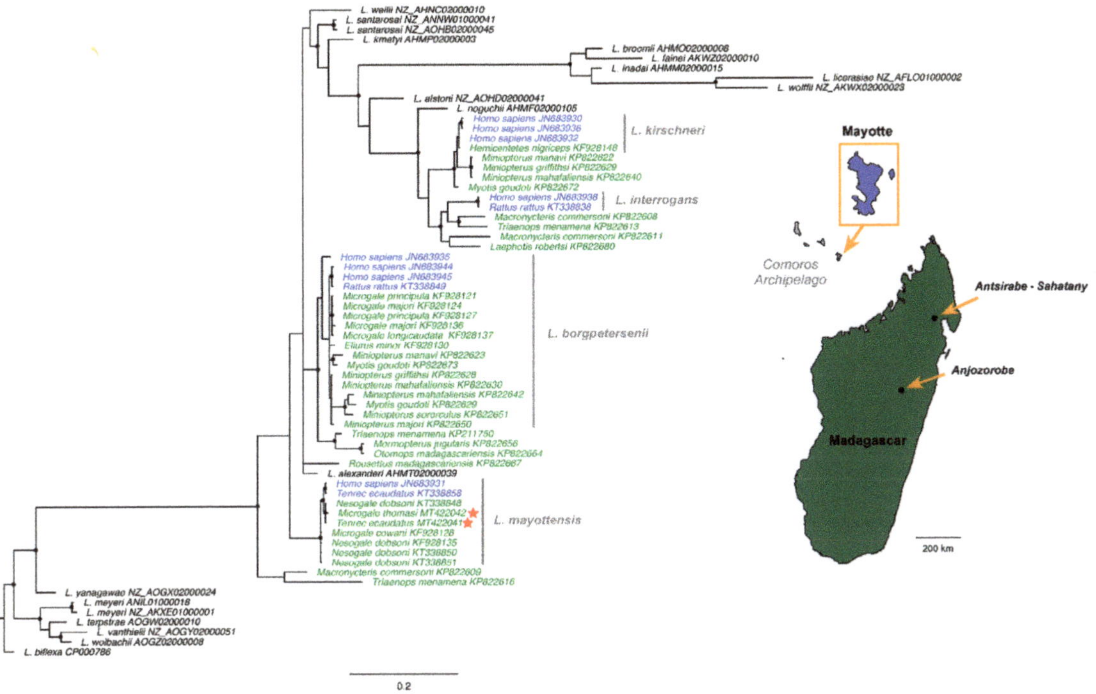

Figure 1. Geographical context and Bayesian phylogenetic tree of *Leptospira* species from Mayotte (blue) and Madagascar (green) based on *secY* gene (482 bp). Sequences in black correspond to *Leptospira* species used as ingroups and outgroup (*L. biflexa*). The accession number is indicated for each sequence. The analysis was conducted under the HKY + I + G substitution model. Black circles at the nodes indicate posterior probabilities superior or equal to 0.90. The red stars indicate new sequences generated in the present study and were obtained from two regions on Madagascar: Anjozorobe and Makira (Commune Antsirabe-Sahatany). The map was realized using worldHires function in mapdata package [27] under the R software version 4.1.1.

2.2. Leptospira Detection and Sequencing

For DNA extraction, kidneys were first rinsed with water and subsequently immersed in 2 mL of sterile water overnight. Then, a thin transversal slice (approximately 0.5 mm thick) was cut in the central part of the kidney using a sterile scalpel, chopped into small pieces, and then submerged into lysis buffer provided in the DNeasy Blood and Tissue Kit (Qiagen, Hilden, Germany) used for DNA extraction. All subsequent extraction steps employed the manufacturer's instructions. *Leptospira* detection was then carried out on 2 µL of eluted DNA using a probe-specific Real-Time Polymerase Chain Reaction system (RT-PCR) targeting a fragment of the 16S rRNA gene [28]. DNA templates leading to positive RT-PCR results were further subjected to end-point PCRs targeting the *secY* and *adk* loci of the MLST scheme#3, as previously described [29]. Amplicons were Sanger sequenced on both strands at GenoScreen (Lille, France) using the same PCR primers. The produced chromatograms were visually edited using Geneious software version 9.0.5 [30].

2.3. Statistical Analyses

Infection prevalence presented herein was compared to figures reported in a previous study (see Table 1 in [20]) using Fisher's exact test, with a significance threshold set at $p < 0.5$.

2.4. Phylogeny

A phylogeny was constructed for the *secY* gene based on the bacterial sequences generated in the present study and previous *secY* sequences from other research in the Malagasy Region [9,15,17,20,24] (Table S1), and different *Leptospira* species were used as ingroups and outgroups. The best model of sequence evolution was determined with jModelTest v.2.1.4 [31]. Phylogenetic reconstruction was performed with MrBayes v.3.2.3 [32]. The analysis consisted of two independent runs of four incrementally heated Metropolis Coupled Markov Chain Monte Carlo (MCMCMC) starting from a random tree. MCMCMC was run for 2 million generations with trees and associated model parameters sampled every 100 generations. The convergence level was validated by an average standard deviation of split frequencies inferior to 0.05. The initial 10% of trees for each run were discarded as burn-in and the consensus phylogeny along with posterior probabilities were obtained from the remaining trees. The resulting Bayesian phylogeny was visualized and annotated with FigTree v.1.4.2 [33].

3. Results and Discussion

The detection by RT-PCR indicates a global leptospiral infection rate of 7.3% (4/55) with bacteria detected in three out of the nine tested tenrecid species: *Microgale taiva* (one positive specimen), *M. thomasi* (two positive specimens), and *Tenrec ecaudatus* (one positive specimen). This overall prevalence is not significantly different from that reported in a previous study carried out in other areas of Madagascar, where 5.6% (12/213) of analyzed tenrecids tested positive for *Leptospira* [20]. The PCR protocols allowed leptospiral sequences to be obtained from the RT-PCR-positive *T. ecaudatus* (*secY*) and from one out of the two RT-PCR-positive *M. thomasi* (*secY* and *adk*). No bacterial sequence was obtained from the second RT-PCR-positive *M. thomasi* or from the RT-PCR-positive *M. taiva*. The three sequences were deposited in GenBank (Accession Numbers MT442041-MT442043).

We present in Figure 1 the Bayesian phylogeny obtained from the *secY* gene. Within this phylogeny, the bacterial sequences obtained from *T. ecaudatus* and *M. thomasi* fall in the *L. mayottensis* clade and form a well-supported subclade with a leptospiral sequence obtained from *Nesogale dobsoni*. This subclade is related to one subclade of *L. mayottensis* detected in humans and tenrecs from Mayotte. All previously reported *Leptospira* sequences from *Microgale* and *Nesogale* species are positioned within two distinct clades: *L. borgpetersenii* (*M. longicaudata*, *M. principula*, and *M. majori*) and *L. mayottensis* (*M. cowani* and *N. dobsoni*). Our results further support this topology with the detection of *L. mayottensis* in *M. thomasi* and Malagasy populations of *T. ecaudatus*.

The Tenrecidae are placental mammals grouped within a monophyletic family endemic to Madagascar and composed of nearly 40 species, including confirmed candidate species [22,23,34]. This highly diversified family is currently considered the result of a single colonization event originating from East Africa that took place between 30 and 56 million years ago, followed by speciation that resulted in an exceptional adaptive radiation [35,36]. Some tenrecids exhibit a number of biological features unique among mammals, such as the ability of hibernating without interbout arousal, partial heterothermy, or elementary echolocation [34,37].

The deep evolutionary history of the Tenrecidae also makes this family suitable for investigating the development of host–parasite interactions. For example, tenrecids host a diversity of Paramyxoviruses, some of which underwent host switches with introduced Muridae rodents [38]. Tenrecidae are known to be hosts of two species of pathogenic *Leptospira*, namely *L. borgpetersenii* and *L. mayottensis* [17,20,24]. While *L. mayottensis* has been identified in tenrecids (on Madagascar and Mayotte) and acute human cases (on

Mayotte), a study on Madagascar reported the presence of *L. mayottensis* in introduced *Rattus rattus*, but only as co-infections with other *Leptospira* species [39]. The strong host-specificity of *L. mayottensis* towards tenrecids was recently tested through experimental infection in which *L. mayottensis* isolated from *T. ecaudatus* failed to colonize the kidneys of *R. norvegicus* [40]. The present study was carried out to (i) further explore the diversity of *L. mayottensis* sheltered by tenrecids and (ii) confirm a previous hypothesis that proposed *L. mayottensis* arrived on Mayotte with the introduction of *T. ecaudatus* for human consumption.

Analyzed samples confirmed tenrecids as being a reservoir of *L. mayottensis* and added *M. thomasi* to the list of animal reservoirs of this pathogenic bacteria. Of particular importance, we report the first characterization of *L. mayottensis* from *T. ecaudatus* on Madagascar. Together with previous data reported on Mayotte [24], the present work supports the introduction of this mammal species to Mayotte being associated with the emergence of a zoonotic human pathogen, *L. mayottensis* on that island. *Tenrec ecaudatus* has also been introduced to other islands in the Malagasy Region with the purpose of providing bush meat, most notably La Réunion, Mauritius, Mahé (Seychelles), and other islands in the Comoros archipelago, but to our knowledge *L. mayottensis* has not been isolated in these non-native *T. ecaudatus* populations or reported in local human inhabitants. The *L. mayottensis* infection prevalence measured in *T. ecaudatus* (4.2%, $n = 24$) is significantly lower than previously reported on Mayotte (27%, $n = 37$; see [24]), while no positive animals have been reported on La Réunion in two independent studies [11,41]. This pattern might be the result of different origins of the *T. ecaudatus* populations introduced to western Indian Ocean islands. On Madagascar, *T. ecaudatus* is found in a range of different forest types and it would be interesting to document the possible phylogeographic structure of these populations and then try to determine the most plausible geographic origin of the populations that have been introduced to other islands in the Malagasy Region. Alternatively, environmental conditions might be more conducive to *L. mayottensis* transmission among *T. ecaudatus* on Mayotte Island than on Madagascar or La Réunion, a hypothesis challenging to test as it requires comprehensive information on *L. mayottensis* biology, including environmental survival of these bacteria in the different geological and climatic contexts.

Data presented herein support *L. mayottensis* being a zoonotic pathogen originating from Madagascar, although we emphasize that overall infection prevalence is low and, hence, preclude any definitive conclusion. However, in addition to the reports of *L. mayottensis* in tenrecs from Madagascar and neighboring Mayotte, it is important to mention that *L. mayottensis* has not been reported outside of western Indian Ocean islands, an area that has considerable species diversity of small mammals including tenrecs and native rodents [42]. We therefore propose that long-term co-radiation processes between Malagasy endemic small mammals and their hosted infectious agents have led to the emergence of endemic microorganisms with zoonotic potential, such as *L. mayottensis*. It has been hypothesized nearly a century ago that the extreme abundance and unbounded dispersal capacities of microorganisms limit endemism, with the exception of some extreme environments, and that biogeographical patterns result from contemporary selective pressures rather than from limited dispersal capacity. This dogma, often referred to as the Baas Becking hypothesis—*"everything is everywhere but the environment selects"* [43]—has been increasingly challenged, but microbial biogeography is still in its infancy [44,45]. The present study supports that host-specificity needs to be considered as a driver of microbial endemism: the dispersal capacities of host-specific microbes are indeed limited by that of their hosts. In other words, when considering host–parasite pairs, the dispersal capacities of hosts drive the biogeographical patterns of their associated microorganisms and may, in the case of strong host–parasite specificity, lead to microbial endemism.

Supplementary Materials: The following supporting information can be downloaded at: https://www.mdpi.com/article/10.3390/tropicalmed7080193/s1, Table S1. Details on *Leptospira secY* sequences from Mayotte and Madagascar used in the present study: host families and species, geographical origins, *Leptospira* species and GenBank accession number. The asterisks indicate sequences generated in the present study.

Author Contributions: P.T., S.M.G. and C.D.G. conceived the study. S.M.G., C.D.G. and V.S. collected the samples in the field. M.T., G.L., M.A.-V. and Y.G. performed laboratory manipulations. Y.G. performed the analysis. P.T., Y.G., S.M.G. and C.D.G. drafted the first version of the manuscript. All authors have read and agreed to the published version of the manuscript.

Funding: The authors declare that no funds, grants, or other support were received during the preparation of this manuscript.

Institutional Review Board Statement: Biological sampling permits were obtained from the Ministry of Environment and Forests and registered under the following references 20/16/MEEMF/SG/DGF/DAPT/SCBT.Re and 85/18/MEEF/SG/DGF/DSAP/SCB.Re.

Informed Consent Statement: Not applicable.

Data Availability Statement: The produced sequences were deposited in GenBank under the accession numbers MT442041, MT442042, and MT442043.

Acknowledgments: We thank the Malagasy authorities (in particular to the former Ministère de l'Environnement, Ecologie et des Forêts, now Ministère de l'Environnement et du Développement Durable) for granting the research permits.

Conflicts of Interest: The authors declare that they have no competing interest.

References

1. Costa, F.; Hagan, J.E.; Calcagno, J.; Kane, M.; Torgerson, P.; Martinez-Silveira, M.S.; Stein, C.; Abela-Ridder, B.; Ko, A.I. Global morbidity and mortality of leptospirosis: A systematic review. *PLoS Negl. Trop. Dis.* **2015**, *9*, e0003898. [CrossRef] [PubMed]

2. Ristow, P.; Bourhy, P.; Kerneis, S.; Schmitt, C.; Prevost, M.-C.; Lilenbaum, W.; Picardeau, M. Biofilm formation by saprophytic and pathogenic leptospires. *Microbiology* **2008**, *154*, 1309–1317. [CrossRef] [PubMed]

3. Adler, B. History of leptospirosis and *Leptospira*. *Curr. Top. Microbiol. Immunol.* **2015**, *387*, 1–9. [CrossRef] [PubMed]

4. Ko, A.I.; Goarant, C.; Picardeau, M. *Leptospira*: The dawn of the molecular genetics era for an emerging zoonotic pathogen. *Nat. Rev. Microbiol.* **2009**, *7*, 736–747. [CrossRef] [PubMed]

5. Masuzawa, T.; Sakakibara, K.; Saito, M.; Hidaka, Y.; Villanueva, S.Y.A.M.; Yanagihara, Y.; Yoshida, S. Characterization of *Leptospira* species isolated from soil collected in Japan. *Microbiol. Immunol.* **2018**, *62*, 55–59. [CrossRef]

6. Thibeaux, R.; Iraola, G.; Ferrés, I.; Bierque, E.; Girault, D.; Soupé-Gilbert, M.-E.; Picardeau, M.; Goarant, C. Deciphering the unexplored *Leptospira* diversity from soils uncovers genomic evolution to virulence. *Microb. Genom.* **2018**, *4*, e000144. [CrossRef]

7. Thibeaux, R.; Girault, D.; Bierque, E.; Soupé-Gilbert, M.-E.; Rettinger, A.; Douyère, A.; Meyer, M.; Iraola, G.; Picardeau, M.; Goarant, C. Biodiversity of environmental *Leptospira*: Improving identification and revisiting the diagnosis. *Front. Microbiol.* **2018**, *9*, 816. [CrossRef]

8. Vincent, A.T.; Schiettekatte, O.; Goarant, C.; Neela, V.K.; Bernet, E.; Thibeaux, R.; Ismail, N.; Khalid, M.K.N.M.; Amran, F.; Masuzawa, T.; et al. Revisiting the taxonomy and evolution of pathogenicity of the genus *Leptospira* through the prism of genomics. *PLoS Negl. Trop. Dis.* **2019**, *13*, e0007270. [CrossRef]

9. Bourhy, P.; Collet, L.; Lernout, T.; Zinini, F.; Hartskeerl, R.A.; van der Linden, H.; Thiberge, J.M.; Diancourt, L.; Brisse, S.; Giry, C.; et al. Human *Leptospira* isolates circulating in Mayotte (Indian Ocean) have unique serological and molecular features. *J. Clin. Microbiol.* **2012**, *50*, 307–311. [CrossRef]

10. Cosson, J.-F.; Picardeau, M.; Mielcarek, M.; Tatard, C.; Chaval, Y.; Suputtamongkol, Y.; Buchy, P.; Jittapalapong, S.; Herbreteau, V.; Morand, S. Epidemiology of *Leptospira* transmitted by rodents in Southeast Asia. *PLoS Negl. Trop. Dis.* **2014**, *8*, e2902. [CrossRef]

11. Guernier, V.; Lagadec, E.; Cordonin, C.; Minter, G.L.; Gomard, Y.; Pagès, F.; Jaffar-Bandjee, M.-C.; Michault, A.; Tortosa, P.; Dellagi, K. Human leptospirosis on Reunion Island, Indian Ocean: Are rodents the (only) ones to blame? *PLoS Negl. Trop. Dis.* **2016**, *10*, e0004733. [CrossRef] [PubMed]

12. Biscornet, L.; Dellagi, K.; Pagès, F.; Bibi, J.; Comarmond, J.D.; Mélade, J.; Govinden, G.; Tirant, M.; Gomard, Y.; Guernier, V.; et al. Human leptospirosis in Seychelles: A prospective study confirms the heavy burden of the disease but suggests that rats are not the main reservoir. *PLoS Negl. Trop. Dis.* **2017**, *11*, e0005831. [CrossRef] [PubMed]

13. Guernier, V.; Richard, V.; Nhan, T.; Rouault, E.; Tessier, A.; Musso, D. *Leptospira* diversity in animals and humans in Tahiti, French Polynesia. *PLoS Negl. Trop. Dis.* **2017**, *11*, e0005676. [CrossRef] [PubMed]

14. Gomard, Y.; Dellagi, K.; Goodman, S.M.; Mavingui, P.; Tortosa, P. Tracking animal reservoirs of pathogenic *Leptospira*: The right test for the right claim. *Trop. Med. Infect. Dis.* **2021**, *6*, 205. [CrossRef]

15. Gomard, Y.; Dietrich, M.; Wieseke, N.; Ramasindrazana, B.; Lagadec, E.; Goodman, S.M.; Dellagi, K.; Tortosa, P. Malagasy bats shelter a considerable genetic diversity of pathogenic *Leptospira* suggesting notable host-specificity patterns. *FEMS Microbiol. Ecol.* **2016**, *92*, fiw037. [CrossRef]

16. Tortosa, P.; Dellagi, K.; Mavingui, P. Leptospiroses on the French Islands of the Indian Ocean. *Bull. Epidemiol. Hebd.* **2017**, *8–9*, 157–161.

17. Dietrich, M.; Gomard, Y.; Lagadec, E.; Ramasindrazana, B.; Le Minter, G.; Guernier, V.; Benlali, A.; Rocamora, G.; Markotter, W.; Goodman, S.M.; et al. Biogeography of *Leptospira* in wild animal communities inhabiting the insular ecosystem of the western Indian Ocean islands and neighboring Africa. *Emerg. Microbes Infect.* **2018**, *7*, 57. [CrossRef]

18. Bourhy, P.; Collet, L.; Clément, S.; Huerre, M.; Ave, P.; Giry, C.; Pettinelli, F.; Picardeau, M. Isolation and characterization of new *Leptospira* genotypes from patients in Mayotte (Indian Ocean). *PLoS Negl. Trop. Dis.* **2010**, *4*, e724. [CrossRef]

19. Bourhy, P.; Collet, L.; Brisse, S.; Picardeau, M. *Leptospira mayottensis* sp. nov., a pathogenic species of the genus *Leptospira* isolated from humans. *Int. J. Syst. Evol. Microbiol.* **2014**, *64*, 4061–4067. [CrossRef] [PubMed]

20. Dietrich, M.; Wilkinson, D.A.; Soarimalala, V.; Goodman, S.M.; Dellagi, K.; Tortosa, P. Diversification of an emerging pathogen in a biodiversity hotspot: *Leptospira* in endemic small mammals of Madagascar. *Mol. Ecol.* **2014**, *23*, 2783–2796. [CrossRef] [PubMed]

21. Dietrich, M.; Gomard, Y.; Cordonin, C.; Tortosa, P. *Leptospira* bacteria in Madagascar. In *The New Natural History of Madagascar*; Goodman, S.M., Ed.; Princeton University Press: Princeton, NJ, USA, 2022; pp. 268–277.

22. Everson, K.M.; Soarimalala, V.; Goodman, S.M.; Olson, L.E. Multiple loci and complete taxonomic sampling resolve the phylogeny and biogeographic history of Tenrecs (Mammalia: Tenrecidae) and reveal higher speciation rates in Madagascar's humid forests. *Syst. Biol.* **2016**, *65*, 890–909. [CrossRef] [PubMed]

23. Goodman, S.M.; Soarimalala, V.; Olson, L.E. Systématique des tenrecs endémiques malgaches (famille des Tenrecidae)/Systematics of endemic Malagasy Tenrecs (family Tenrecidae). In *Les Aires Protégées Terrestres de Madagascar: Leur Histoire, Description, et Biote./The Terrestrial Protected Areas of Madagascar: Their History, Description and Biota*; Goodman, S.M., Raherilalao, M.J., Wohlauser, S., Eds.; Association Vahatra: Antananarivo, Madagascar, 2018; pp. 363–372.

24. Lagadec, E.; Gomard, Y.; Minter, G.L.; Cordonin, C.; Cardinale, E.; Ramasindrazana, B.; Dietrich, M.; Goodman, S.M.; Tortosa, P.; Dellagi, K. Identification of *Tenrec ecaudatus*, a wild mammal introduced to Mayotte Island, as a reservoir of the newly identified human pathogenic *Leptospira* mayottensis. *PLoS Negl. Trop. Dis.* **2016**, *10*, e0004933. [CrossRef] [PubMed]

25. Annapragada, A.; Brook, C.E.; Luskin, M.S.; Rahariniaina, R.P.; Helin, M.; Razafinarivo, O.; Ambinintsoa Ralaiarison, R.; Randriamady, H.J.; Olson, L.E.; Goodman, S.M.; et al. Evaluation of tenrec population viability and potential sustainable management under hunting pressure in northeastern Madagascar. *Anim. Conserv.* **2021**, *24*, 1059–1070. [CrossRef]

26. Sikes, R.S. The Animal Care and Use Committee of the American Society of Mammalogists. Guidelines of the American Society of Mammalogists for the use of wild mammals in research and education. *J. Mammal.* **2016**, *97*, 663–688. [CrossRef]

27. Becker, R.A.; Wilks, A.R. R Version by Ray Brownrigg. Mapdata: Extra Map Databases. R Package Version 2.3.0. 2018. Available online: https://cran.r-project.org/web/packages/mapdata/ (accessed on 12 August 2022).

28. Smythe, L.D.; Smith, I.L.; Smith, G.A.; Dohnt, M.F.; Symonds, M.L.; Barnett, L.J.; McKay, D.B. A quantitative PCR (TaqMan) assay for pathogenic *Leptospira* spp. *BMC Infect. Dis.* **2002**, *2*, 13. [CrossRef]

29. Ahmed, N.; Devi, S.M.; Valverde, M.D.L.A.; Vijayachari, P.; Machang'u, R.S.; Ellis, W.A.; Hartskeerl, R.A. Multilocus sequence typing method for identification and genotypic classification of pathogenic *Leptospira* species. *Ann. Clin. Microbiol. Antimicrob.* **2006**, *5*, 28. [CrossRef] [PubMed]

30. Kearse, M.; Moir, R.; Wilson, A.; Stones-Havas, S.; Cheung, M.; Sturrock, S.; Buxton, S.; Cooper, A.; Markowitz, S.; Duran, C.; et al. Geneious basic: An integrated and extendable desktop software platform for the organization and analysis of sequence data. *Bioinformatics* **2012**, *28*, 1647–1649. [CrossRef]

31. Darriba, D.; Taboada, G.L.; Doallo, R.; Posada, D. jModelTest 2: More models, new heuristics and parallel computing. *Nat. Methods* **2012**, *9*, 772. [CrossRef]

32. Ronquist, F.; Teslenko, M.; van der Mark, P.; Ayres, D.L.; Darling, A.; Höhna, S.; Larget, B.; Liu, L.; Suchard, M.A.; Huelsenbeck, J.P. MrBayes 3.2: Efficient Bayesian phylogenetic inference and model choice across a large model space. *Syst. Biol.* **2012**, *61*, 539–542. [CrossRef]

33. Rambaut, A. FigTree 1.4.2, a Graphical Viewer of Phylogenetic Trees. 2014. Available online: http://tree.bio.ed.ac.uk/software/figtree/ (accessed on 12 August 2022).

34. Olson, L.E. Tenrecs. *Curr. Biol.* **2013**, *23*, R5–R8. [CrossRef]

35. Emerson, B.C. Evolution on oceanic islands: Molecular phylogenetic approaches to understanding pattern and process. *Mol. Ecol.* **2002**, *11*, 951–966. [CrossRef] [PubMed]

36. Krause, D.W. Washed up in Madagascar. *Nature* **2010**, *463*, 613–614. [CrossRef] [PubMed]

37. Lovegrove, B.G.; Lobban, K.D.; Levesque, D.L. Mammal survival at the Cretaceous–Palaeogene boundary: Metabolic homeostasis in prolonged tropical hibernation in tenrecs. *Proc. Biol. Sci.* **2014**, *281*, 20141304. [CrossRef] [PubMed]

38. Wilkinson, D.A.; Mélade, J.; Dietrich, M.; Ramasindrazana, B.; Soarimalala, V.; Lagadec, E.; Minter, G.L.; Tortosa, P.; Heraud, J.-M.; Lamballerie, X.D.; et al. Highly diverse Morbillivirus-related paramyxoviruses in wild fauna of the southwestern Indian Ocean islands: Evidence of exchange between introduced and endemic small mammals. *J. Virol.* **2014**, *88*, 8268–8277. [CrossRef]

39. Moseley, M.; Rahelinirina, S.; Rajerison, M.; Garin, B.; Piertney, S.; Telfer, S. Mixed *Leptospira* infections in a diverse reservoir host community, Madagascar, 2013–2015. *Emerg. Infect. Dis.* **2018**, *24*, 1138–1140. [CrossRef]

40. Cordonin, C.; Turpin, M.; Bringart, M.; Bascands, J.-L.; Flores, O.; Dellagi, K.; Mavingui, P.; Roche, M.; Tortosa, P. Pathogenic *Leptospira* and their animal reservoirs: Testing host specificity through experimental infection. *Sci. Rep.* **2020**, *10*, 7239. [CrossRef]
41. Desvars, A.; Naze, F.; Benneveau, A.; Cardinale, E.; Michault, A. Endemicity of leptospirosis in domestic and wild animal species from Reunion Island (Indian Ocean). *Epidemiol. Infect.* **2013**, *141*, 1154–1165. [CrossRef]
42. Goodman, S.M.; Soarimalala, V. Introduction to mammals. In *The New Natural History of Madagascar*; Goodman, S.M., Ed.; Princeton University Press: Princeton, NJ, USA, 2022; pp. 1737–1769.
43. Baas Becking, L.G.M. *Geobiologie of Inleiding Tot de Milieukunde*; W.P. Van Stockum & Zoon NV: The Hague, The Netherlands, 1934.
44. De Wit, R.; Bouvier, T. 'Everything is everywhere, but, the environment Selects'; what did Baas Becking and Beijerinck really say? *Environ. Microbiol.* **2006**, *8*, 755–758. [CrossRef]
45. Fierer, N. Microbial biogeography: Patterns in microbial diversity across space and time. In *Accessing Uncultivated Microorganisms: From the Environment to Organisms and Genomes and Back*; Zengler, K., Ed.; ASM Press: Washington, WA, USA, 2008; pp. 95–115.

Tropical Medicine and Infectious Disease

Article

Pathological Abnormalities Observed on Ultrasonography among Fishermen Associated with Male Genital Schistosomiasis (MGS) along the South Lake Malawi Shoreline in Mangochi District, Malawi

Sekeleghe A. Kayuni [1,2,*], Mohammad H. Al-Harbi [1], Peter Makaula [3], Boniface Injesi [2], Bright Mainga [4], Fanuel Lampiao [5], Lazarus Juziwelo [6], E. James LaCourse [1] and J. Russell Stothard [1]

[1] Department of Tropical Disease Biology, Liverpool School of Tropical Medicine, Liverpool L3 5QA, UK
[2] MASM Medi Clinics Limited, Medical Society of Malawi (MASM), Lilongwe P.O. Box 1254, Malawi
[3] Research for Health, Environment and Development (RHED), Mangochi P.O. Box 345, Malawi
[4] Laboratory Department, Mangochi District Hospital, Mangochi District Assembly, Mangochi P.O. Box 1854, Malawi
[5] Physiology Department, School of Life Sciences and Allied Health Professions, Kamuzu University of Health Sciences, Mahatma Gandhi Road, Blantyre 312225, Malawi
[6] National Schistosomiasis and Soil-Transmitted Helminths Control Programme, Community Health Sciences Unit, Ministry of Health, Lilongwe P.O. Box 30377, Malawi
* Correspondence: sekekayuni@live.com; Tel.: +265-888-367367

Citation: Kayuni, S.A.; Al-Harbi, M.H.; Makaula, P.; Injesi, B.; Mainga, B.; Lampiao, F.; Juziwelo, L.; LaCourse, E.J.; Stothard, J.R. Pathological Abnormalities Observed on Ultrasonography among Fishermen Associated with Male Genital Schistosomiasis (MGS) along the South Lake Malawi Shoreline in Mangochi District, Malawi. *Trop. Med. Infect. Dis.* **2022**, *7*, 169. https://doi.org/10.3390/tropicalmed7080169

Academic Editors: Vyacheslav Yurchenko and Thomas Dorlo

Received: 7 July 2022
Accepted: 29 July 2022
Published: 5 August 2022

Publisher's Note: MDPI stays neutral with regard to jurisdictional claims in published maps and institutional affiliations.

Abstract: Schistosome eggs cause granulomata and pathological abnormalities, detectable with non-invasive radiological techniques such as ultrasonography which could be useful in male genital schistosomiasis (MGS). As part of our novel MGS study among fishermen along Lake Malawi, we describe pathologies observed on ultrasonography and praziquantel (PZQ) treatment over time. Fishermen aged 18+ years were recruited, submitted urine and semen for parasitological and molecular testing, and thereafter, transabdominal pelvic and scrotal ultrasonography, assessing pathologies in the prostate, seminal vesicles, epididymis and testes. Standard PZQ treatment and follow-up invitation at 1-, 3-, 6- and 12-months' time-points were offered. A total of 130 recruited fishermen underwent ultrasonography at baseline (median age: 32.0 years); 27 (20.9%, $n = 129$) had *S. haematobium* eggs in urine (median: 1.0 egg/10 mL), 10 (12.3%, $n = 81$) in semen (defined as MGS, median: 2.9 eggs/mL ejaculate) and 16 (28.1%, $n = 57$) had a positive seminal *Schistosoma* real-time PCR. At baseline, 9 fishermen (6.9%, $n = 130$) had abnormalities, with 2 positive MGS having prostatic and testicular nodules. Fewer abnormalities were observed on follow-up. In conclusion, pathologies detected in male genitalia by ultrasonography can describe MGS morbidity in those with positive parasitological and molecular findings. Ultrasonography advances and accessibility in endemic areas can support monitoring of pathologies' resolution after treatment.

Keywords: MGS; ultrasonography; prostate; epididymis; testis; seminal vesicles

1. Introduction

Schistosomiasis is a prevalent parasitic disease in SSA where more than 90% of infected people live, causing considerable morbidity and some deaths [1,2]. Male genital schistosomiasis (MGS) is a specific-gender manifestation of urogenital schistosomiasis (UGS), associated with schistosome eggs and related pathologies in genitalia of men inhabiting or visiting endemic areas in SSA [3,4]. Despite the first recognised report described by Madden in 1911 [5], MGS remains poorly recognised and described owing to limited research over several decades.

Schistosome eggs evoke immunological responses causing granulomata formation and pathological lesions which give characteristic manifestations in the urinary tract and the liver, detectable by non-invasive radiological techniques such as ultrasonography which has become widely available in endemic areas [6]. This technique is safe, effective and can be a valuable diagnostic method in tropical diseases, such as the control of NTDs [7,8].

Ultrasonography is now portable and can be easily moved from one health facility to another, potentially supporting the semen microscopy and other available point-of-care diagnostics tests for schistosomiasis.

Transrectal ultrasonography (TRUS), computed tomography (CT) and magnetic resonance imaging (MRI) have been noted to be useful in schistosomiasis of urinary tract and liver [9]. These methods detect lesions concomitant with granulomata and calcifications, can determine pathology of various organs, as well as demonstrate calcified schistosome ova in the tissues [10–13].

Such pathologies have also been anecdotally observed to be correlated with clinical features specific to MGS and noted to resolve in earlier stages compared to irreversible, late presentation [14,15]. Furthermore, Niamey ultrasonography criteria was developed by World Health Organisation (WHO, Geneva, Switzerland) to describe the pathological lesions in renal tract and liver, associated with urogenital schistosomiasis (UGS) as well as intestinal schistosomiasis [8]. Currently there is very limited evidence in literature from few case reports of scrotal/lower abdominal studies in travellers describing MGS and impact on ultrasound pathologies of PZQ in an endemic setting, which necessitate for further studies to validate the potential use of ultrasonography in MGS and understand the evolution and resolution of the associated genital pathologies.

As part of our novel longitudinal cohort study on MGS among fishermen in south-western shoreline of Lake Malawi, known to harbour both urogenital and intestinal schistosome species [16,17], a study was conducted to describe the pathological abnormalities observed on ultrasonography over time, which could be attributed to MGS and effects of PZQ treatment(s).

2. Materials and Methods

2.1. Study Area, Population and Sampling

The research study was conducted among fishermen living in fishing communities (villages) identified and selected along the southern shoreline of Lake Malawi in Mangochi district, the largest district in southern region of Malawi, from October 2017 to December 2018 (Figure 1). Fishermen aged $\geq$ 18 years willing to provide written informed consent were eligible to participate in the study, as described in the earlier publications [18,19].

2.2. Study Data Collection

The study methods used for the data collection included individual questionnaires, parasitological analyses of urine and semen samples collected in health facilities [18,19]. In addition, all participants were invited to undergo transabdominal pelvic and scrotal ultrasonography examination using a portable Chison Q5 ultrasound scanner with 3.5 MHz probe (Mount International United Services Ltd., Gloucester, UK) to assess pathological abnormalities in the prostate, seminal vesicles, testes and epididymis (Figure 2). The endorectal probe which is the standard of care for such examinations was considered too invasive for this study and would not be widely available outside a study context.

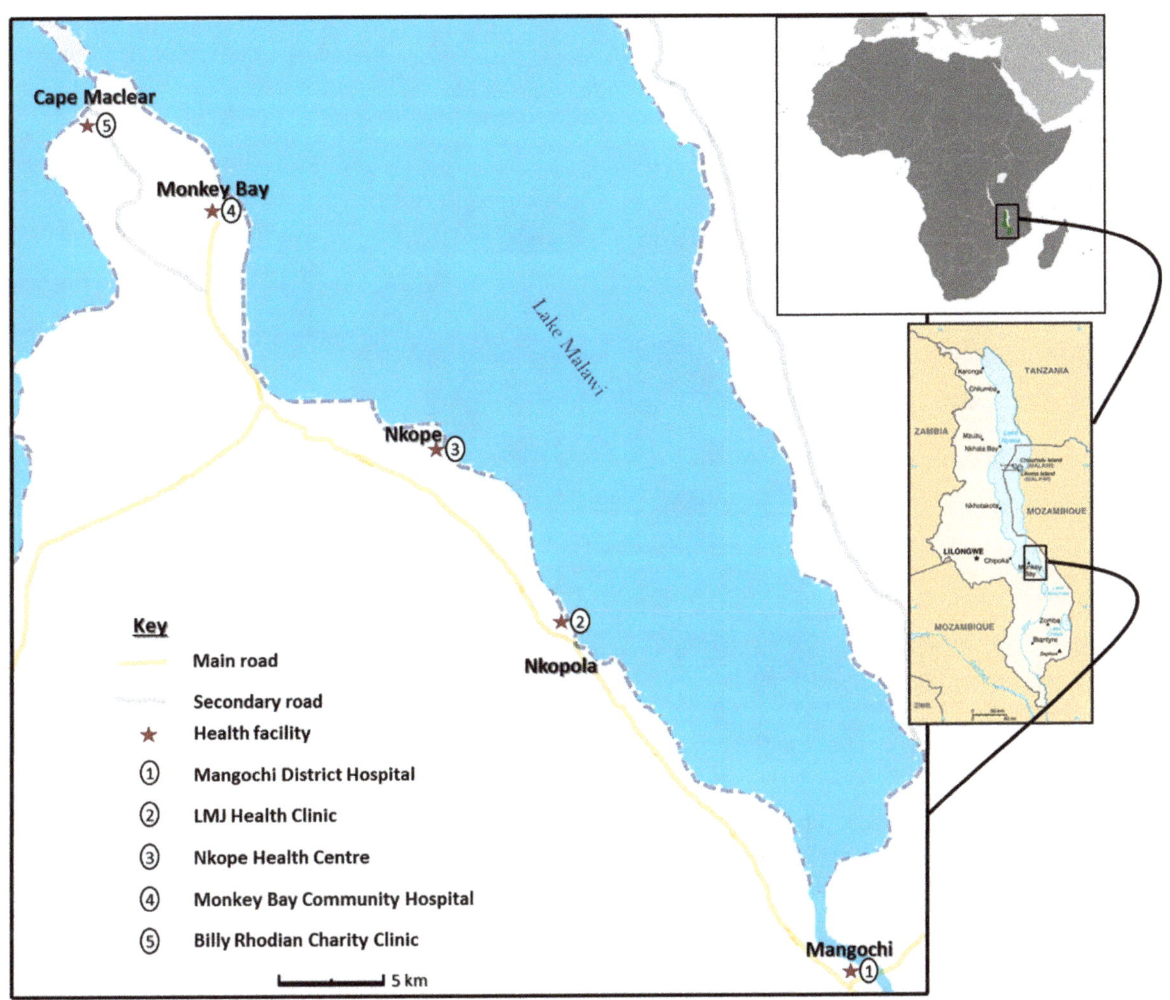

Figure 1. Schematic map of Study area showing health facilities along Lake Malawi. (The study map was produced by Sekeleghe Kayuni (on 4 August 2019), while the maps of Africa and Malawi were reproduced from the maps at the Central Intelligence Agency (CIA) website, public domain: https://www.cia.gov/library/publications/the-world-factbook/attachments/locator-maps/MI-locator-map.gif accessed on 4 August 2019 and https://www.cia.gov/library/publications/the-world-factbook/attachments/maps/MI-map.gif accessed on 4 August 2019).

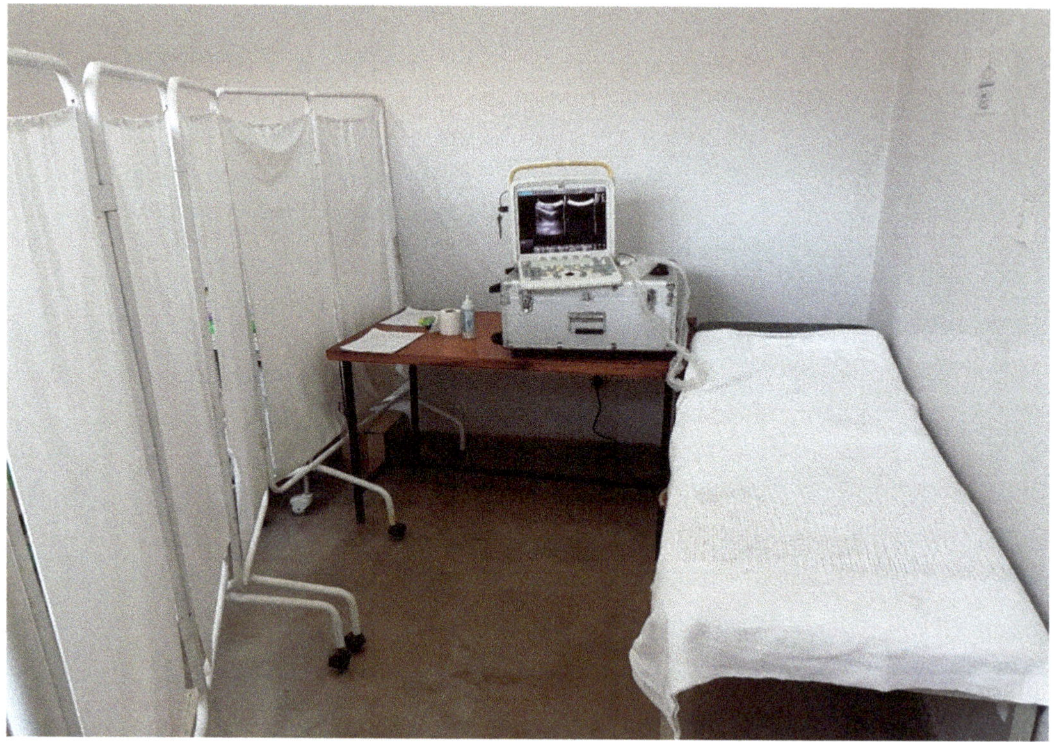

Figure 2. The portable Chison Q5 ultrasound scanner in an examination room.

2.3. Ultrasonography Examination of Urogenital Organs

2.3.1. Preparations for the Procedure

Safety precautions including use of appropriate protective wear and gloves were ensured during the ultrasonography examinations. Participants were briefed on the transabdominal and scrotal ultrasonography procedures. The scanner was set on the urology exam mode for the procedure. Participants were asked to present with a full bladder, before the procedure to increase the quality of the images. Whenever possible, room lightning was turned off to maximise screen visibility.

2.3.2. Outline of the Ultrasonographic Procedure

The participant's study number was registered in the ultrasound machine and report form prior to commencing the procedure. The participant was positioned supine on the examination couch. The scanning procedure investigated the urinary bladder, seminal vesicles and scrotum (testis, epididymis). Image quality was recorded first and then absence/presence of pathological findings were documented.

Urinary Bladder and Kidneys

Transverse (TS) and longitudinal (LS)sweeps through the bladder were performed to assess the shape (distension) and wall thickness, as well as the distal ureters where possible.

Schistosomiasis-related bladder pathologies included a rounded or irregular shape of the bladder, wall thickening with diffused or focal thickening of >5 mm (mild: 6–10 mm; severe: $\geq$11 mm), bladder wall calcifications and masses or pseudopolyps protruding in the bladder lumen; the distal ureters were considered pathological when dilated.

After performing several sweeps through the bladder, the best representative sweep was stored as a video. Bladder wall thickness was measured in mm and stored as a separate still image. In case of any pathologic findings, additional still images with relevant measurements were stored. If the bladder wall thickness was abnormal, the kidneys were scanned for evidence of hydronephrosis.

Prostate

The prostate was visualised during scanning of the bladder. Normal volume was set at 30 mm^3 or less with smooth outline. Pathological findings potentially consistent with of schistosomiasis included nodules or masses larger than 1 cm, and calcifications of the prostate. After performing several sweeps through the prostate, the best representative sweep was stored under the label "prostate". In case of any pathologic findings, additional still images with relevant measurements were stored.

Seminal Vesicles

Seminal vesicles were scanned in the TS plane. Normal appearances were defined as the seminal vesicles being symmetrical and measuring 15 mm or less in antero-posterior (AP) dimension with a smooth outline. Pathological findings potentially consistent with schistosomiasis were defined as enlarged and/or asymmetrical vesicles with a nodular, hyperechoic appearance.

Storage of images and clips: if the vesicles measured larger than 15 mm in AP plane, their measurement were stored as a separate still image. Following performing several sweeps through the vesicles, the best representative sweep was stored under the label "SV". In case of any pathologic findings, additional still images with relevant measurements were stored.

Scrotum

Transverse sweeps of the scrotum were performed to assess both testes.

Testis abnormalities potentially suggestive of schistosomiasis were defined as nodules or calcifications; any other abnormalities of testis, epididymis and scrotum were also documented.

2.4. Disinfection and Patient Information after the Completion of the Procedure

At the end of the procedure the probe was cleaned with tissue paper to remove the gel, and with methylated spirit. All participants were notified of pathological findings that day by the study clinician, and further appropriate investigations and management were organised in accordance with the standard clinical practice. Thereafter, praziquantel treatment at 40 mg/kg as a single dose was offered along with an invitation to follow-up studies after 1-, 3-, 6- and 12-months.

2.5. Statistical Analyses

All the information collected during the study was screened and quality-controlled before entry into Microsoft Excel and SSPS computer packages. No double data entry was conducted. Screening for errors using descriptive analyses and cleaning were conducted, before commencing statistical analyses to present the results of the study. All video clips and digital images were coded for data protection and were then stored onto the device before transferring to a password-protected external hard drive for further analyses. A sample of 15% of the scan images was randomly selected and re-read by specialist radiologist for quality control, who conducted training on ultrasonography of urogenital organs.

All video clips and digital images collected from ultrasonography were stored onto the device before transferring to the external hard drive for further analyses. The data collected from the clips, images and report forms were screened to clear all errors before entry into IBM SSPS programmes in line with previous findings on specific ultrasonography [8,13,20]. Summary statistics were calculated to explore the data and thereafter correlations and significant tests were conducted to describe and interpret the results further, mainly using nonparametric tests.

2.6. Ethical Considerations

Ethical clearance to conduct the study was provided by the National Health Sciences Research Committee (NHSRC, approval number: 1805) of Malawi and Liverpool School of Tropical Medicine (LSTM) Research Ethics Committee (LSTM REC, approval number: 17-018), as outlined earlier. Utmost privacy and confidentiality were maintained in the study and where necessary, the information was anonymised to protect the identity of the participant.

Since this was a test-and-treat study, participants were notified of the ultrasonography results including pathological findings at the end of the procedure and where necessary, further appropriate investigations and management were organised in accordance to the standard clinical practice. Treatment with PZQ at 40 mg/kg as a single dose was offered before inviting them to the next follow-up studies at 1-month, 3-, 6- and 12-months' time-points. Details of observed treatment were recorded when subsequent follow-ups were performed.

3. Results

Of the 376 fishermen recruited into the study from 39 villages located in two Traditional Authorities (T/A) of Mponda and Nankumba, only 130 participants returned to the health facility for the ultrasonography examinations at baseline of the study.

3.1. Demographic Information and Diagnostic Results

The median age of the 130 scanned participants was 32.0 years with a range of 19.0 to 70.0 years (Interquartile range [IQR]: 18) and their duration of stay in the fishing village ranged from 2 months to 70 years (median: 22.0; IQR: 24.5; Table 1). The median weight of the participants was 59.0 kg (IQR: 9.0, range: 43.0–75.4 kg).

Table 1. Demographical information, diagnostic analyses on urine and semen results of the 130 study participants.

Variable	*n*	Median	Range	Interquartile Range (IQR)	Number of Positive Cases	Prevalence (%)
Age	130	32.0	19.0–70.0	18.0	-	-
Duration of stay in village (years)	120	22.0	0.2–70.0	24.5	-	-
Weight (kgs)	115	59.0	43.0–75.4	9.0	-	-
Eggs in urine (filtration, 10 mL)	129	1.0	0.1–186.0	5.8	27	20.9
POC-CCA	129	-	-	-	6	4.9
Eggs in semen (mL)	81	2.9	0.4–9.3	4.6	10	12.3
Seminal real-time PCR (Ct-value)	57	26.4	18.9–36.6	10.5	16	28.1%

All participants except one submitted urine and 81 submitted semen (62.3%). After urine filtration, 27 participants (20.9%) had *S. haematobium* eggs in urine (UGS), their mean egg count was 19.1 eggs per 10 mL and ranging from 0.1 to 186.0 eggs (median: 1.0, IQR: 5.8). Six participants (4.9%) had a positive POC-CCA test, suggestive of possible intestinal *S. mansoni* infection.

For the 81 participants who submitted semen, 10 (12.3%) had *S. haematobium* eggs in semen (MGS), mean egg count was 3.9 per ml of ejaculate (median: 2.9 eggs), ranging from 0.4 to 9.3 eggs and volume of semen ranged from 0.1 to 4.5 mL (mean: 1.6 mL). The real-time PCR conducted on 57 semen samples revealed that 16 participants (28.1%) were positive. Four participants were positive on both semen microscopy and real-time PCR.

3.2. Baseline Results of the Ultrasonography Examinations

Of the participants who had ultrasonography, 9 (6.9%) participants had abnormalities in genital organs. Specifically, abnormalities were noted in prostate, seminal vesicles and/or scrotum (testis and epididymis) (Table 2). One participant who had abnormalities in urinary bladder wall with severe polypoid thickness was detected to have bilateral hydronephrosis (Table 3).

Table 2. Proportion of abnormal findings in particular organs at baseline.

Organ	Total Scans	Number of Abnormal Scans	Percentage (%)
Urinary Bladder *	106	2	1.9%
Prostate	126	3	2.4%
Seminal vesicles	117	1	0.9%
Testis [#]	129	1	0.8%
Epididymis [‡]	129	1	0.8%
Scrotum [α]	129	6	4.7%

* Bilateral hydronephrosis in 1 participant, [#] left testis; [‡] right epididymis; [α] hydroceles were observed in scrotums of six participants.

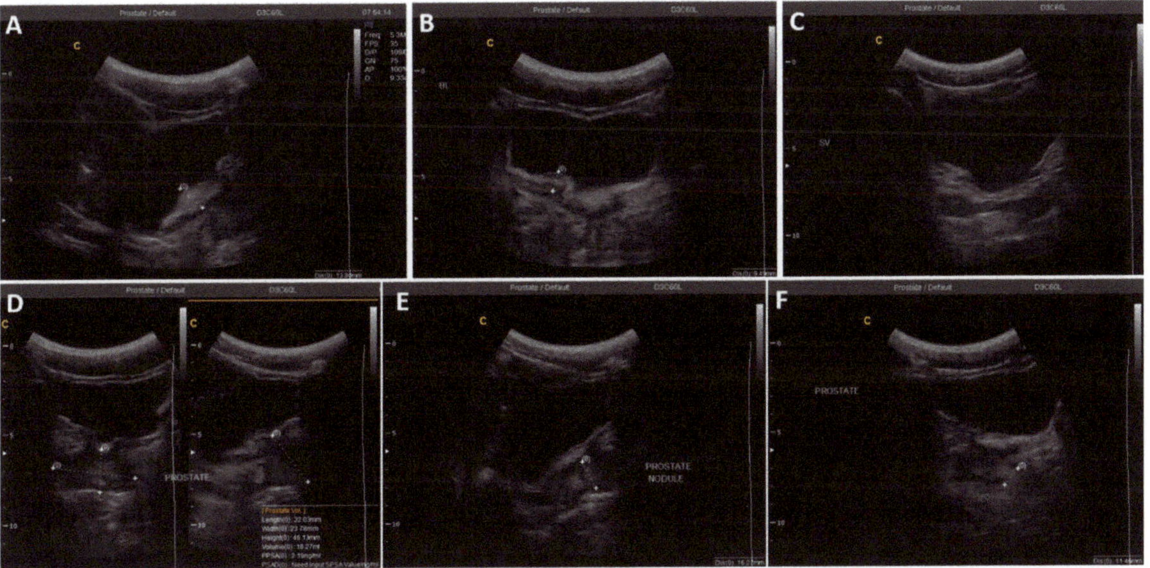

Figure 3. Ultrasonographic images of an MGS positive study participant with abnormalities in the bladder and prostate at Baseline. (**A,B**). Irregular urinary bladder wall and severe focal thickness, measuring up to 13.6 mm. (**C**). Normal symmetrical seminal vesicles. (**D**). Prostate with abnormal irregular outline, but normal volume of 18.3 mL. (**E,F**). Hyperechoic nodule in the prostate, measuring 11.4 mm by 16.2 mm.

Eighteen of all the scanned participants had UGS only, confirmed by *S. haematobium* eggs observed in urine, 15 had MGS only (six had semen eggs only while 9 were positive for real-time PCR) and only two participants were positive for all three diagnostic tests. Only 4 participants with sonographic abnormalities had MGS confirmed by semen microscopy and real-time PCR, two participants had schistosome eggs in urine and semen and positive real-time PCR.

Of the four participants with abnormalities in at least 2 GU organs as shown in Table 3 below, the abnormalities of 2 with confirmed MGS could be classical description of its morbidity (participants 2 and 3):

Table 3. Abnormalities observed in participants with at least 2 GU organs affected at baseline.

Participant	Age (Years)	Eggs in Urine (per 10 mL)	Eggs in Semen (per mL)	Real-Time PCR (Ct-Value)	Abnormalities Observed
1	19	0	0	N/D [#]	Irregular bladder wall and severe polypoid thickness, with bilateral hydronephrosis
2	22	0	0	25.4	Irregular bladder wall with severe focal thickness, irregular prostate with hyperechoic nodule (Figure 3)
3	49	1	6	N/D [#]	Left testicular nodule and mild bilateral hydroceles
4	69	0	N/A [‡]	N/A [‡]	Severely enlarged prostate (volume = 61.3 mL) and right epididymis, with bilateral hydrocele

[#] Test not done, inadequate sample; [‡] sample not submitted, test not done.

3.2.1. Urinary Bladder and Kidneys

Two participants with abnormal ultrasound findings had irregular outline of their urinary bladder with severe wall thickness of at least 11 mm, one had bilateral hydronephrosis (participant 1, Table 3) while the other (participant 2, Table 3) had severe focal bladder thickness.

3.2.2. Prostate

Three participants (2.3%) had abnormal prostate appearances, two aged 51 and 69 years old, had enlarged prostates with volumes of 39.1 mL and 61.3 mL. The other participant, aged 22 years with confirmed MGS through positive semen real-time PCR, had irregular prostate outline and hyperechoic nodule (Figure 3), attributable to description of MGS morbidity. A point-of-care prostate specific antigen (POC-PSA) conducted on these participants was negative, excluding possibility of prostatic infection, hyperplasia or tumour (Figure 4).

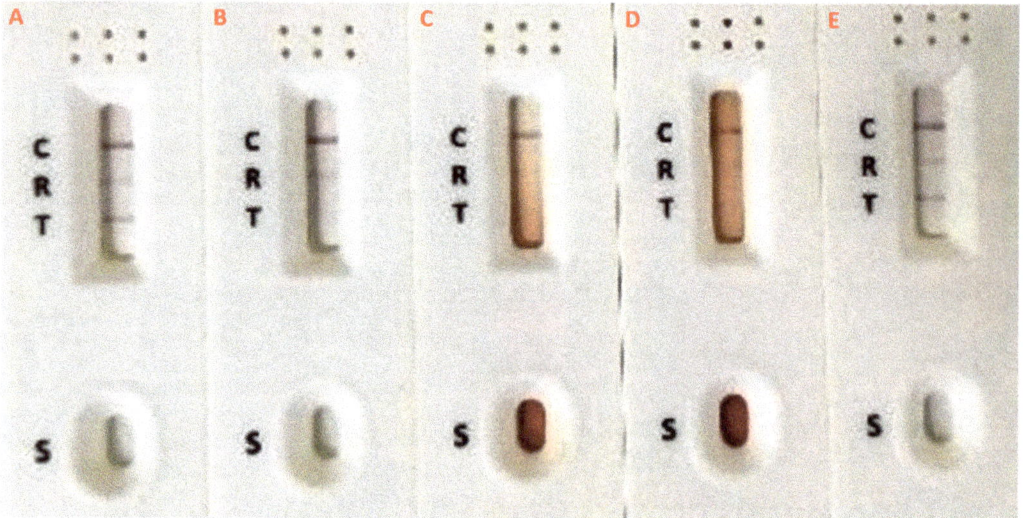

Figure 4. Images of the POC-PSA test conducted on participants with Prostate abnormalities. (**A,E**). Strong positive POC-PSA with no urine or semen eggs and no abnormalities. (**B**). Negative POC-PSA with semen real-time PCR (Ct-value: 25.4), irregular prostate outline and hyperechoic nodule. (**C**). Negative POC-PSA with no urine eggs, but had grossly irregular, enlarged prostate (61.3 mL), abnormal epididymis and bilateral hydrocele. (**D**). Negative POC-PSA with no urine or semen eggs, negative real-time PCR but had irregular, enlarged prostate (39.1 mL).

3.2.3. Seminal Vesicles

Only one participant had asymmetrical, hyperechoic vesicles, suggestive of MGS related pathology, aged 24 years with *S. haematobium* eggs in urine and semen, as well as positive real-time PCR. The measurements for the scanned participants' right vesicle were from 4.8 mm to 18.5 mm and the left vesicle from 3.7 mm to 17.5 mm.

3.2.4. Scrotum

One participant (0.8%) had a left testicular nodule, with *S. haematobium* eggs detected in urine and semen (Figure 5). No testicular abnormalities were detected in the other MGS participants. Only one participant, aged 69, was observed to have an abnormal enlarged right epididymis (participant 4, Table 3), which could be attributable to MGS morbidity.

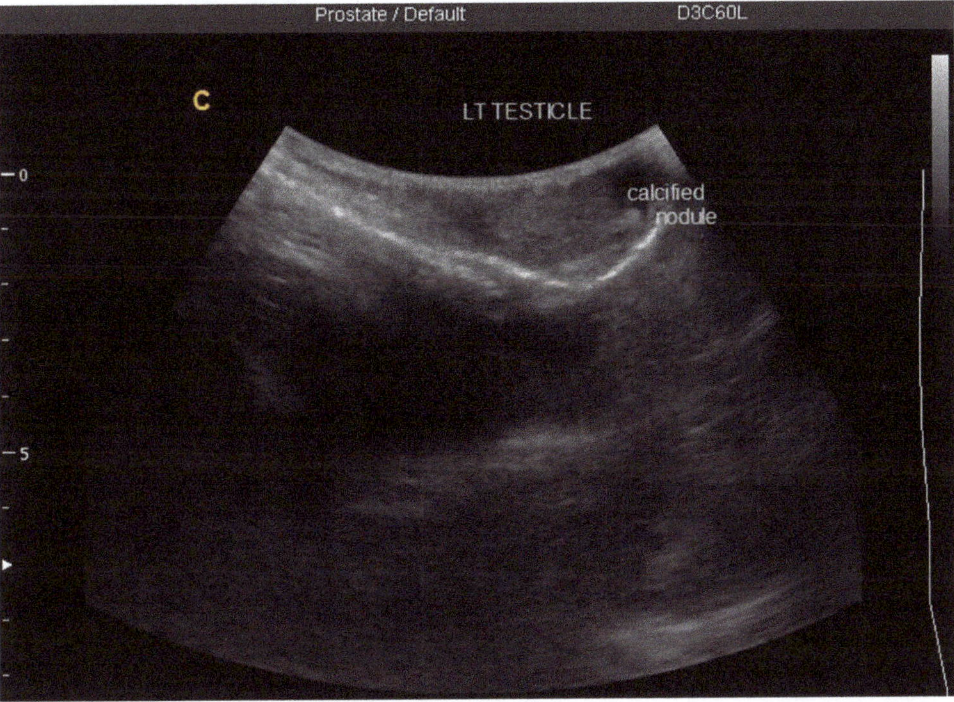

Figure 5. Ultrasonographic image of study participant with left testicular nodule at baseline.

In context with all the abnormalities described above, there was no correlation between age, duration of stay, diagnostic tests' results and abnormalities.

3.3. Follow-Up Ultrasonography Examinations

At the end of the ultrasonography examinations at baseline, PZQ treatment was provided to the participants on exit of the study after submitting semen sample for resolution of the abnormalities. The participants were invited to follow-up ultrasonography examinations at 1-, 3-, 6- and 12 months' time-points.

3.3.1. One-Month Follow-Up

Only 29 participants were scanned out of the 60 participants who returned at 1-month follow-up, with four participants scanned for the first time and no abnormalities observed. Only one of two participants with classical description of MGS morbidity was scanned at this time-point and had bilateral hydroceles despite PZQ treatment and negative tests (Table 4).

Table 4. Observations on ultrasonography of one participant at baseline and one-month follow-up.

Age (Years)	Baseline		1-Month Follow-Up	
	Test Results	**Abnormalities Observed**	**Test Results**	**Abnormalities Observed**
49	Eggs in urine, semen; no real-time PCR done	Left testicular nodule, mild bilateral hydroceles	No eggs in urine or semen; negative PCR	Bilateral hydroceles

3.3.2. Three-Months Follow-Up

Sixty-four participants were followed up at 3-months' time-point of which 32 had ultrasonography examinations, and 4 were scanned for the first time. On diagnostic examinations, 5 had *S. haematobium* eggs in urine (17.2%), 4 in semen (13.8%), 4 had trace POC-CCA test while 5 had positive semen real-time PCR.

The MGS participant with prostate nodule at baseline had no abnormality detected while the testicular nodule participant was lost-to-follow-up. Three other participants (28.1%) had hydroceles at this time-point.

3.3.3. Six-Months Follow-Up

Sixty-three participants were followed up at 6-months' time-point of which 38 had ultrasonography examinations, and 4 were scanned for the first time. On diagnostic examinations, 2 had *S. haematobium* eggs in urine (5.3%), 1 in semen (2.6%), 4 (10.5%) positive and 2 (5.3%) trace POC-CCA test results, while 3 had positive semen real-time PCR (7.6%).

The MGS participant with prostatic nodule at MGS was lost to follow-up, while a 33-year-old participant with negative diagnostic tests had enlarged seminal vesicles and other three had hydroceles.

3.3.4. Twelve-Months Follow-Up

Forty-five participants were reviewed at 12-months' time-point of which 17 had ultrasonography examinations, and 4 were scanned for the first time. One participant had abnormal bladder wall thickness and left kidney mass. On diagnostic examinations, the participants were negative for POC-CCA, urine filtration and semen microscopy, with 2 participants being positive on semen real-time PCR (11.8%).

Table A1 (in Appendix A) illustrates the progress of abnormalities detected at baseline, over the course of the study and Table A2 shows those with no abnormality at baseline but detected during the follow-up. In total, 146 participants were scanned in the study and abnormalities were noted in 16 participants at various time points (Table A3).

4. Discussion

To our knowledge, this is the first prospective ultrasonographic study of MGS in Malawi and South-eastern Africa on the fishermen cohort to determine its morbidity, through observations of pathological abnormalities in genital organs, as well as look at changes through time in men after standard dose-regimen of PZQ treatment. Our study observed classical abnormalities which could describe the morbidity of MGS, namely prostatic and testicular nodules in two confirmed participants on parasitological and molecular testing.

4.1. Genital Consequences of Schistosomiasis on Ultrasonography

Genital manifestations such as MGS are among the complications of schistosomiasis which remains unknown among local inhabitants frequently exposed to infective cercariae harbouring their freshwater bodies as well as health professionals working in the areas which result in undiagnosed, under- or mistreating and underreporting of the disease, contributing further to morbidity among affected men. In some instances, people suffer from social prejudice and discrimination arising from the consequences of the genital complications, such as infertility, abnormal organ swelling, reduced sexual prowess, coital pain, genital bleeding among other with women severely and disproportionately [21,22].

In order to improve awareness and knowledge of MGS, other diagnostic methods can be added to help in detection of the disease. Radiological techniques have been observed to improve diagnosis of schistosomiasis of liver and urological tract at various stages, mostly with chronic complications. Ultrasonography is considered as an acceptable, safe and less-invasive tool in diagnosis, management and monitoring control of NTDs, including schistosomiasis. Recent advances in this technology have resulted in development of portable, high quality scanning devices which can be easily mobile to limited-resource endemic areas and utilised in the available infrastructure, in detecting the genital pathological abnormalities affecting rural population among other conditions, as demonstrated in our study.

4.2. Pathological Abnormalities Associated with MGS in Malawian Fishermen

Our MGS cohort study among local fishermen along the south shoreline of Lake Malawi observed a 17.1% baseline prevalence of UGS, 10.4% for MGS using semen microscopy and 26.6% by semen real-time PCR. Among those 130 participants who were scanned at baseline, the prevalence of MGS was 12.3% using semen microscopy and 28.1% by semen real-time PCR.

Previous studies have described abnormalities observed in genital organs such as prostate, seminal vesicles, ejaculatory ducts, vas deferens, epididymis, tests, scrotal sac among other structures, which include organ enlargement, shrinkage, dilatations, thickening, echogenic lesions, calcification, hydroceles among others, which can mimic other diseases. In schistosomiasis-endemic areas, detection of such pathological abnormalities together with classical genital symptoms such as genital, coital or ejaculatory pain, haemospermia, abnormal ejaculates, reduced libido or suspected infertility, could suggests a diagnosis of MGS [19]. Since similar clinical presentation and findings can be associated with other diseases prevalent in these endemic areas, affecting the genital organs such as sexually transmitted infections (STIs), tuberculosis (TB), malignant hypertension or cancer, proper clinical assessment and extensive diagnostic examinations are very important to be conducted to ensure appropriate diagnosis, treatment, care and management of affected men.

The findings of this study showed that ten participants had pathological abnormalities in their GU organs at baseline, of which nine participants (6.9%) having them in prostate, seminal vesicles and scrotum and two had classical abnormalities, descriptive of MGS morbidity. These abnormalities suggest consequences of previous or current schistosomal infection acquired from their frequent exposure during fishing and other routine activities in the lake, which is known to harbour schistosomes. Seminal vesicles were observed to have abnormalities in the study, consistent with evidence from previous studies and literature which describes that seminal vesicles are among the frequent affected genital organs with schistosomiasis [15,23,24].

Further results show that urinary bladder had abnormal wall thickening, in some cases severe polypoid structures and associated bilateral hydronephrosis which required referral for further medical management at district hospital, unfortunately, participant was lost to follow-up. Such lesions have been widely described in the literature originating from schistosome worms which have matured as male and female worms in hepatic venules, pair up and migrate to the vesical plexus where they reside around urinary bladder. These worms continually deposit massive number of schistosome eggs, which get trapped in the bladder wall and cause granulomatous reactions, fibrosis, calcifications and architecture destruction, thereby compromising bladder functioning [25,26]. As the infection progresses and bladder malfunctions, urine backflow compromises the ureters (hydroureter) which later affects the kidneys, resulting in hydronephrosis. Early diagnosis and management are critical in preventing such chronic and fatal consequences of schistosomiasis.

Interestingly, schistosome worms have been thought to reside in venous plexus around the genital organs such as prostate, seminal vesicle and testes, with eggs being trapped in the tissues due to its tough architecture in comparison to the urinary bladder. This can result in echogenic lesions, calcifications, organ enlargement, atrophy, hydroceles among the pathological abnormalities in these genital organs which can be detected on ultrasonography [11,14,27]. These abnormalities were observed in this study in the prostate, epididymis and testis, with some participants being positive on the urine and semen diagnostic tests. Prostate abnormalities were observed in three participants, of which two had grossly enlarged prostates and one had hyperechoic prostatic nodule and positive semen real-time PCR. These can be attributed to MGS after exclusion of other possible diseases such as STIs, TB, benign prostatic hyperplasia or prostate malignancy. In addition, the lack of statistically significant differences among those with abnormalities in accordance to age, duration of stay and diagnostic tests' results demonstrate that the lesions can present at any age since they could have developed from a young age, as reported previously [28,29].

4.3. Pathological Abnormalities after Treatment

Monitoring of disease morbidity especially MGS is critical in controlling the disease and prevention of severe irreversible pathological abnormalities which later could contribute to mortality. As a mainstay treatment, PZQ has shown to be effective in treating both forms of schistosomiasis, registering cure rates of over 90% in most endemic areas [30,31]. Currently, it is utilised by most national control programmes in endemic areas as one of the key control interventions through the MDA campaigns, which commonly targeted school-aged children. PZQ has also been used in treating MGS, clearing the schistosome eggs in semen and resolving some pathological abnormalities, while adjusting the standard dose of treatment in some cases to ensure complete cure [15].

From the study follow-up after PZQ treatment to the participants, pathological abnormalities were not detectable in most participants on follow-ups. This could illustrate the knowledge that early mild abnormalities will be resolved by standard PZQ treatment as it kills those adult-laying worms, hence reducing further damage to the organs. Other chronic, long-standing abnormalities such as hydroceles detected on follow-up, required further assessments, medical and surgical interventions to resolve these, as it was done in the study where such participants were referred to the bigger district hospital.

Repeated exposure to the infested lake water can contribute to newer abnormalities developing after PZQ treatment. Moreover, longer duration in resolution of abnormalities support the need for repeated PZQ treatment to completely resolve the severe abnormalities which are reversible, while providing additional control interventions such as adequate awareness and health education, provision of adequate, portable, safe and clean water, encouraging construction and utilisation of household and community sanitation facilities, as well environmental control to reduce intermediate snail host population.

4.4. Study Limitations and Ultrasonographic Diagnostic Challenges in MGS

The low number of participants undergoing the ultrasonography examinations and inadequate volumes of semen samples to run real-time PCR limit the tests results' comparisons and generalisation of the study results to male population in endemic region. This could be explained by lack of experience with the method as most participants reported this was their first time to undergo such examinations. Negative perceptions with regards to new techniques in rural communities and the longer time spent during the examination especially among those presenting with inadequate bladder filling could have deterred more study participants from taking part. Moreover, some participants could be reluctant to take part at health centres, due to poor health-seeking behaviour. More sensitisation and discussions which were done can help to address such and other concerns, thereby stimulate more participants to such important studies.

The low sensitivity of transabdominal ultrasonography compared to TRUS, CT or MRI could result in missing some lesions in the genital organs, resulting in poorly described burden of genital diseases such as MGS. However, cost implications associated with these sensitive techniques, their unavailability and inaccessibility as well as low acceptability among local participants could further jeopardise the implementations of such examinations in rural endemic areas.

In addition, advanced radiological expertise and extensive training on genital ultrasonography are required in order to detect pathological abnormalities arising from MGS, which could be easily mistaken for those from other prevalent genital diseases such as STIs, TB or cancer. Moreover, further ultrasonography studies with larger cohort as well as inclusion of more advanced portable devices such as TRUS are necessary in endemic areas to describe more on morbidity of MGS.

5. Conclusions

In conclusion, pathological abnormalities can be detected using portable transabdominal and scrotal ultrasonography, as demonstrated in our study, which together with positive parasitological and molecular MGS tests could describe its morbidity. Owing to advances in portable ultrasonography and their accessibility in endemic areas especially SSA, together with symptomatology description and parasitological findings, this diagnostic technique can also aid in monitoring in MGS treatment and resolution of its related pathologies.

Author Contributions: Conceptualisation, S.A.K., E.J.L. and J.R.S.; methodology, formal analysis, investigation, resources and data curation, S.A.K., M.H.A.-H., P.M., B.I., B.M., F.L., L.J., E.J.L. and J.R.S.; writing—original draft preparation, review and editing, S.A.K. All authors have read and agreed to the published version of the manuscript.

Funding: This research was funded by a PhD scholarship for S.A.K. from Commonwealth Scholarship Commission, UK, an International Travel Fellowship from the British Society for Parasitology, grant from World Friendship Charity and African Research Network for Neglected Tropical Diseases (ARNTD) through United States Agency for International Development (USAID), UK aid from the British people (UKaid) and Coalition for Operational Research on Neglected Tropical Diseases (COR-NTD). The contents of this publication are solely the responsibility of the authors and not the funders.

Institutional Review Board Statement: The study was conducted in accordance with the Declaration of Helsinki and approved by the National Health Sciences Research Committee (NHSRC, approval number: 1805) of Malawi and Liverpool School of Tropical Medicine (LSTM) Research Ethics Committee (LSTM REC, approval number: 17-018).

Informed Consent Statement: Written informed consent was obtained from all participants recruited into the study for participation and publication of its findings.

Data Availability Statement: The data for this study has been presented within this article and any further information regarding this study can be reasonably requested from the corresponding author.

Acknowledgments: Many thanks to the Director of Health and Social Services, District Medical Officers and management team of Mangochi District Health Office for their overwhelming support toward the study; Sam Kampondeni and Elizabeth Joekes for their assistance in the training on ultrasonography; and Joekes for her assistance in quality assurance of the sonographic images; Jutta Reinhard-Rupp and Merck plc for donation of the portable ultrasound unit; Messrs Pilirani Mkambeni, Patrick Hussein, Mkonazi Nkhoma and Matthews Elias for technical assistance; and all the fishermen who participated in the longitudinal cohort study in Mangochi district. We are also grateful to the head and staff of the laboratory department at Mangochi District Hospital; district ART coordinators, in-charge, and staff of Tikondane clinic; in-charges and staff of Billy Riordan Memorial Clinic; Monkey-Bay Community Hospital; Nkope and Koche Health Centres; Lumbua Katenda, Anwar, and staff of Mangochi LMJ clinic; local community health workers: Ambali Makochera, Alfred Kachikowa, Rodgers Wengawenga, Michael Tsatawe, Chikondi Mtsindula, Mambo Amin, Elias Matemba, Austin Kaluwa, Justin Mndala, Promise Mwawa, Dickson Tabu, George Matiki, Alfred Mdoka, and Charles Katandi; Brother Henry Chagoma and staff of Montfort Mission Guest

House; and all traditional leaders in the hosting fishing communities and beach committee members for their enthusiasm and support during the study.

Conflicts of Interest: The authors declare no conflict of interest. The funders had no role in the design of the study; in the collection, analyses, or interpretation of data; in the writing of the manuscript, or in the decision to publish the results.

Appendix A

Table A1. Ten participants with abnormalities in Pelvic organs (9 in genital organs) on ultrasonography at baseline and progression over the time points in the study.

Age (Years)	Baseline	1-Month	3-Months	6-Months	12-Months
19	Bladder thickening, severe hydronephrosis	Lost to follow-up			
22	Bladder thickening, prostate nodule	Did not show up	No abnormality	Inadequate bladder filling, not reported for repeat scan	Lost to follow-up
24	Enlarged asymmetrical seminal vesicles	Inadequate bladder filling, not reported for repeat scan	Lost to follow-up		
29	Left hydrocele	Lost to follow-up			
30	Left hydrocele	Did not show up	Did not show up	Did not show up	No abnormality
49	Bilateral hydroceles	Bilateral hydroceles	Bilateral hydroceles	Bilateral hydroceles	Lost to follow-up
49	Testicular nodule, mild bilateral hydroceles	Bilateral hydroceles	Lost to follow-up		
51	Enlarged prostate	Did not show up	Did not show up	No abnormality	Lost to follow-up
54	Bilateral hydroceles	Lost to follow-up			
69	Enlarged prostate, bilateral hydroceles	Lost to follow-up			

Table A2. Six participants with abnormalities in GU organs on ultrasonography during follow-up time-points but none at baseline in the study.

Age (Years)	Baseline	1-Month	3-Months	6-Months	12-Months
25	No abnormality	No abnormality	*Did not show up*	Left hydrocele	Lost to follow-up
33	No abnormality	*Did not show up*	Left hydrocele	*Did not show up*	No abnormality
33	No abnormality	*Did not show up*	*Did not show up*	Enlarged, asymmetrical seminal vesicles	Lost to follow-up
39	No abnormality	*Did not show up*	*Did not show up*	Left hydrocele	Lost to follow-up
44	No abnormality	Left hydrocele	Left hydrocele	No abnormality	Lost to follow-up
53	No abnormality	*Did not show up*	*Did not show up*	*Did not show up*	Bladder thickening, left kidney mass

Table A3. Proportion of participants in the study with abnormalities in GU organs on ultrasonography in accordance with their MGS status at all time-points.

		MGS									
		Baseline (*n* = 130)		1-Month (*n* = 29)		3-Months (*n* = 32)		6-Months (*n* = 38)		12-Months (*n* = 17)	
		Positive	Negative	Positive	Negative	Positive	Negative	Positive	Negative	Positive	Negative
USS	Positive	4	5	2	1	1	2	0	4	0	0
	Negative	14	107	1	25	3	26	1	33	2	15

References

1. McManus, D.P.; Dunne, D.W.; Sacko, M.; Utzinger, J.; Vennervald, B.J.; Zhou, X.-N. Schistosomiasis. *Nat. Rev. Dis. Primers* **2018**, *4*, 13. [CrossRef] [PubMed]
2. WHO. Schistosomiasis Geneva, Switzerland: World Health Organization. 2018. Available online: http://www.who.int/en/news-room/fact-sheets/detail/schistosomiasis (accessed on 20 February 2018).
3. Feldmeier, H.; Leutscher, P.; Poggensee, G.; Harms, G. Male genital schistosomiasis and haemospermia. *Trop. Med. Int. Health* **1999**, *2*, 791–793. [CrossRef] [PubMed]
4. Leutscher, P.; Ramarokoto, C.-E.; Reimert, C.; Feldmeier, H.; Esterre, P.; Vennervald, B.J. Community-based study of genital schistosomiasis in men from Madagascar. *Lancet* **2000**, *355*, 117–118. [CrossRef]
5. Madden, F.C. Two rare manifestations of Bilharziosis. *Lancet* **1911**, *178*, 754–755. [CrossRef]
6. WHO. *Manual of Diagnostic Ultrasound*, 2nd ed.; World Health Organization and World Federation for Ultrasound in Medicine and Radiology: Geneva, Switzerland, 2011.
7. Ramarakoto, C.E.; Leutscher, P.D.C.; van Dam, G.; Christensen, N.O. Ultrasonographical findings in the urogenital organs in women and men infected with *Schistosoma haematobium* in northern Madagascar. *Trans. R. Soc. Trop. Med. Hyg.* **2008**, *102*, 767–773. [CrossRef]
8. WHO. *Ultrasound in Schistosomiasis: A Practical Guide to the Standard Use of Ultrasonography for Assessment of Schistosomiasis-Related Morbidity: Second International Workshop, October 22–26 1996, Niamey, Niger*; WHO: Geneva, Switzerland, 2000.
9. Shebel, H.M.; Elsayes, K.M.; Abou El Atta, H.M.; Elguindy, Y.M.; El-Diasty, T.A. Genitourinary schistosomiasis: Life cycle and radiologic-pathologic findings. *Radiogr. Rev. Publ. Radiol. Soc. N. Am. Inc.* **2012**, *32*, 1031–1046. [CrossRef]
10. Fender, D.; Hamdy, F.C.; Neal, D.E. Transrectal ultrasound appearances of schistosomal prostato-seminovesiculitis. *Br. J. Urol.* **1996**, *77*, 166–167. [CrossRef]
11. Al-Saeed, O.; Sheikh, M.; Kehinde, E.O.; Makar, R. Seminal vesicle masses detected incidentally during transrectal sonographic examination of the prostate. *J. Clin. Ultrasound JCU* **2003**, *31*, 201–206. [CrossRef]
12. de Cassio Saito, O.; de Barros, N.; Chammas, M.C.; Oliveira, I.R.S.; Cerri, G.G. Ultrasound of tropical and infectious diseases that affect the scrotum. *Ultrasound Q.* **2004**, *20*, 12–18. [CrossRef]
13. Vilana, R.; Corachan, M.; Gascon, J.; Valls, E.; Bru, C. Schistosomiasis of the male genital tract: Transrectal sonographic findings. *J. Urol.* **1997**, *158*, 1491–1493. [CrossRef]
14. Richter, J. Evolution of schistosomiasis-induced pathology after therapy and interruption of exposure to schistosomes: A review of ultrasonographic studies. *Acta Trop.* **2000**, *77*, 111–131. [CrossRef]
15. Kayuni, S.; Lampiao, F.; Makaula, P.; Juziwelo, L.; Lacourse, E.J.; Reinhard-Rupp, J.; Leutscher, P.D.C.; Stothard, J.R. A systematic review with epidemiological update of male genital schistosomiasis (MGS): A call for integrated case management across the health system in sub-Saharan Africa. *Parasite Epidemiol. Control* **2019**, *4*, e00077. [CrossRef] [PubMed]
16. Alharbi, M.H.; Condemine, C.; Christiansen, R.; LaCourse, E.J.; Makaula, P.; Stanton, M.C.; Juziwelo, L.; Kayuni, S.; Stothard, J.R. *Biomphalaria pfeifferi* snails and intestinal schistosomiasis, Lake Malawi, Africa, 2017–2018. *Emerg. Infect. Dis.* **2019**, *25*, 613–615. [CrossRef] [PubMed]
17. Makaula, P.; Sadalaki, J.R.; Muula, A.S.; Kayuni, S.; Jemu, S.; Bloch, P. Schistosomiasis in Malawi: A systematic review. *Parasites Vectors* **2014**, *7*, 570. [CrossRef] [PubMed]
18. Kayuni, S.A.; Corstjens, P.L.A.M.; Lacourse, E.J.; Bartlett, K.E.; Fawcett, J.; Shaw, A.; Makaula, P.; Lampiao, F.; Juziwelo, L.; de Dood, C.J.; et al. How can schistosome circulating antigen assays be best applied for diagnosing male genital schistosomiasis (MGS): An appraisal using exemplar MGS cases from a longitudinal cohort study among fishermen on the south shoreline of Lake Malawi. *Parasitology* **2019**, *146*, 1785–1795. [CrossRef]
19. Kayuni, S.A.; LaCourse, E.J.; Makaula, P.; Lampiao, F.; Juziwelo, L.; Fawcett, J.; Shaw, A.; Alharbi, M.H.; Verweij, J.J.; Stothard, J.R. Case report: Highlighting male genital schistosomiasis (MGS) in fishermen from the southwestern shoreline of Lake Malawi, Mangochi District. *Am. J. Trop. Med. Hyg.* **2019**, *101*, 1331–1335. [CrossRef]
20. Martino, P.; Galosi, A.B.; Bitelli, M.; Consonni, P.; Fiorini, F.; Granata, A.; Gunelli, R.; Liguori, G.; Palazzo, S.; Pavan, N.; et al. Practical recommendations for performing ultrasound scanning in the urological and andrological fields. *Arch. Ital. Urol. Androl.* **2014**, *86*, 56–78. [CrossRef]
21. Kukula, V.A.; MacPherson, E.E.; Tsey, I.H.; Stothard, J.R.; Theobald, S.; Gyapong, M. A major hurdle in the elimination of urogenital schistosomiasis revealed: Identifying key gaps in knowledge and understanding of female genital schistosomiasis within communities and local health workers. *PLoS Negl. Trop. Dis.* **2019**, *13*, e0007207. [CrossRef]
22. Yirenya-Tawiah, D.R.; Ackumey, M.M.; Bosompem, K.M. Knowledge and awareness of genital involvement and reproductive health consequences of urogenital schistosomiasis in endemic communities in Ghana: A cross-sectional study. *Reprod. Health* **2016**, *13*, 117. [CrossRef]
23. Alves, W.; Woods, R.; Gelfand, M. The distribution of bilharzia ova in the male genital tract. *Cent. Afr. J. Med.* **1955**, *1*, 166–168.
24. Gelfand, M.; Ross, C.M.D.; Blair, D.M.; Castle, W.M.; Webber, M.C. Schistosomiasis of the male pelvic organs: Severity of infection as determined by digestion of tissue and histologic methods in 300 cadavers. *Am. J. Trop. Med. Hyg.* **1970**, *19*, 779–784. [CrossRef] [PubMed]
25. Bustinduy, A.L.; King, C.H. Schistosomiasis. In *Manson's Tropical Diseases*, 23rd ed.; Farrar, J., Hotez, P.J., Junghanss, T., Kang, G., Lalloo, D., White, N.J., Eds.; Elsevier Health Sciences: Beijing, China, 2014; pp. 698–725.
26. Colley, D.G.; Bustinduy, A.L.; Secor, W.E.; King, C.H. Human schistosomiasis. *Lancet* **2014**, *383*, 2253–2264. [CrossRef]

27. Richens, J. Genital manifestations of tropical diseases. *Sex. Transm. Infect.* **2004**, *80*, 12–17. [CrossRef] [PubMed]
28. Ekenze, S.O.; Modekwe, V.O.; Nzegwu, M.A.; Ekpemo, S.C.; Ezomike, U.O. Testicular schistosomiasis mimicking malignancy in a child: A case report. *J. Trop. Pediatrics* **2015**, *61*, 304–309. [CrossRef]
29. Rambau, P.F.; Chandika, A.; Chalya, P.L.; Jackson, K. Scrotal swelling and testicular atrophy due to schistosomiasis in a 9-year-old boy: A case report. *Case Rep. Infect. Dis.* **2011**, *2011*, 787961. [CrossRef]
30. Knopp, S.; Becker, S.L.; Ingram, K.J.; Keiser, J.; Utzinger, J. Diagnosis and treatment of schistosomiasis in children in the era of intensified control. *Expert Rev. Anti-Infect.* **2013**, *11*, 1237–1258. [CrossRef]
31. WHO. *Assessing the Efficacy of Anthelminthic Drugs against Schistosomiasis and Soil-Transmitted Helminthiases*; World Health Organization: Geneva, Switzerland, 2013; p. 29.

Tropical Medicine and Infectious Disease

MDPI

Review

Establishing and Integrating a Female Genital Schistosomiasis Control Programme into the Existing Health Care System

Takalani Girly Nemungadi [1,2,*], Tsakani Ernica Furumele [2], Mary Kay Gugerty [3], Amadou Garba Djirmay [4], Saloshni Naidoo [1] and Eyrun Flörecke Kjetland [1,5]

1   Discipline of Public Health Medicine, School of Nursing and Public Health, College of Health Sciences, University of KwaZulu-Natal, Durban 4000, South Africa
2   Communicable Disease Control Directorate, National Department of Health, Pretoria 0001, South Africa
3   Evans School of Public Policy & Governance, University of Washington, Seattle, WA 98195-3055, USA
4   Department of the Control of Neglected Tropical Diseases, World Health Organization, 1211 Geneva, Switzerland
5   Norwegian Centre for Imported and Tropical Diseases, Department of Infectious Diseases Ullevaal, Oslo University Hospital, 0424 Oslo, Norway
*   Correspondence: takalaninemungadi@gmail.com

Citation: Nemungadi, T.G.; Furumele, T.E.; Gugerty, M.K.; Djirmay, A.G.; Naidoo, S.; Kjetland, E.F. Establishing and Integrating a Female Genital Schistosomiasis Control Programme into the Existing Health Care System. *Trop. Med. Infect. Dis.* **2022**, 7, 382. https://doi.org/10.3390/tropicalmed7110382

Academic Editors: Vyacheslav Yurchenko and Thomas Dorlo

Received: 29 September 2022
Accepted: 9 November 2022
Published: 16 November 2022

Publisher's Note: MDPI stays neutral with regard to jurisdictional claims in published maps and institutional affiliations.

Abstract: Female genital schistosomiasis (FGS) is a complication of *Schistosoma haematobium* infection, and imposes a health burden whose magnitude is not fully explored. It is estimated that up to 56 million women in sub-Saharan Africa have FGS, and almost 20 million more cases will occur in the next decade unless infected girls are treated. Schistosomiasis is reported throughout the year in South Africa in areas known to be endemic, but there is no control programme. We analyze five actions for both a better understanding of the burden of FGS and reducing its prevalence in Africa, namely: (1) schistosomiasis prevention by establishing a formal control programme and increasing access to treatment, (2) introducing FGS screening, (3) providing knowledge to health care workers and communities, (4) vector control, and (5) water, sanitation, and hygiene. Schistosomiasis is focal in South Africa, with most localities moderately affected (prevalence between 10% and 50%), and some pockets that are high risk (more than 50% prevalence). However, in order to progress towards elimination, the five actions are yet to be implemented in addition to the current (and only) control strategy of case-by-case treatment. The main challenge that South Africa faces is a lack of access to WHO-accredited donated medication for mass drug administration. The establishment of a formal and funded programme would address these issues and begin the implementation of the recommended actions.

Keywords: female genital schistosomiasis (FGS); schistosomiasis; homogenous yellow patch; grainy sandy patch; health care system; South Africa

1. Introduction

Genital grainy sandy patches and homogenous yellow sandy patches are among the gynaecological manifestations of female genital schistosomiasis, and are caused by the presence of *Schistosoma haematobium* ova in genital tissue [1]. *Schistosoma haematobium* is a blood fluke that causes disease, and if not treated, causes subsequent pathology in the urinary tract and genital tissue, the latter called female genital schistosomiasis (FGS) [1]. The global prevalence of FGS is not known, but it has been reported to be high in poor and rural communities in the tropical and subtropical parts of the world, especially those which do not have access to adequate sanitation and safe water [2,3]. It is estimated that up to 56 million women in sub-Saharan Africa have FGS, and almost 20 million more cases will occur in the next decade unless girls are treated [4]. A study in the KwaZulu-Natal province of South Africa, found one or more of the three well-known genital mucosal manifestations of FGS [5]. Of the 2008 to 2009 biopsy-diagnosed schistosomiasis cases in the Limpopo province of South Africa, 87.6% were FGS (n = 233/266) [6].

Anti-schistosomal treatment, before the development of genital lesions, forms part of the prevention of FGS [7]. Treatment before the age of 13 years offers the best protection against *S. haematobium*-induced pathology in the genital tract. However, a single dose of anti-schistosomal treatment does not seem to prevent all genital morbidity [8]. The risk of FGS in Africa is underestimated, schistosomiasis treatment is not universally available, and FGS does not appear anywhere in the health programmes [9–11]. Schistosomiasis is, however, reported throughout the year in focal areas, mostly where there is no continuous access to piped water. In non-endemic areas, a large number of imported cases are reported through the notifiable disease systems of South Africa [12,13]. In rural KwaZulu-Natal, South Africa, high prevalences of FGS, pregnancy, HIV, and sexually transmitted infection (STI) are reported among sexually active young women [14,15].

The burden of FGS is relative to genitourinary schistosomiasis in endemic areas [16,17]. In clinical practice, FGS is misdiagnosed as an STI or cervical cancer, resulting in unnecessary surgery and unwarranted administration of antibiotics [3,11,15,18–23]. Misdiagnosed and wrongly treated individuals therefore continue to experience FGS symptoms and have heightened risks of HIV and HPV infections [24,25]. The lack of knowledge of FGS among clinicians is exacerbated by the neglect of schistosomiasis control [17].

2. The Case of South Africa

2.1. Geography

South Africa is divided into nine provinces: the Eastern Cape, which has two metropolitan municipalities and six district municipalities; the Free State, which has one metropolitan municipality and four district municipalities; Gauteng, which has three metropolitan municipalities and two district municipalities; KwaZulu-Natal, which has one metropolitan municipality and ten district municipalities; Limpopo, which has five district municipalities; Mpumalanga, which has three district municipalities; the Northern Cape, which has five district municipalities; North West, which has four district municipalities; and Western Cape, with one metropolitan municipality and five district municipalities. The provinces vary greatly in size, with Gauteng being the smallest, most densely populated, and most urbanized. Northern Cape is the largest and least populated province.

South Africa has three capitals: Cape Town, in the City of Cape Town metropolitan municipality, Western Cape, is the legislative capital; Bloemfontein, in Mangaung metropolitan municipality, Free State, is the judicial capital; and Pretoria, in the City of Tshwane metropolitan municipality, Gauteng, is the administrative capital, and the ultimate capital of the country.

2.2. Schistosomiasis Prevalence

In South Africa, both recent mappings of schistosomiasis and research on FGS among school learners from various districts and schools shed light on challenges caused by schistosomiasis requiring serious attention (Table 1).

Table 1. Schistosomiasis prevalence studies in South Africa.

Year of Study	Study Area	Province	Number of Schools (Total Number of Participants)	Prevalence of *S. haematobium*	Prevalence of *S. mansoni*	Mean Intensity of *S. haematobium* Infection (Eggs/10 mL Urine)
2009	Ugu District [9]	KwaZulu-Natal	5 (Not reported (NR))	≥50%	NR	NR
2009–2010	Ugu District [26]	KwaZulu-Natal	15 (726) [a]	36.9%	0%	20 (range from 1–624)
2009–2010	Ugu District [7]	KwaZulu-Natal	18 (970)	32%	0%	52
2010	Ugu District [27]	KwaZulu-Natal	18 (1057)	32%	0%	52
2010 2011	Ugu District [b,f] [13]	KwaZulu-Natal	18 (108) 18 (76)	38% and 29.6% [c] 13.2% [f] and 11.8% [c]	NR	18.4 [b] and 14.9 [c] 9 [b] and 12.9 [c]

Table 1. *Cont.*

Year of Study	Study Area	Province	Number of Schools (Total Number of Participants)	Prevalence of *S. haematobium*	Prevalence of *S. mansoni*	Mean Intensity of *S. haematobium* Infection (Eggs/ 10 mL Urine)
2011	Ugu District [9]	KwaZulu-Natal	9 (NR)–	10–49%	NR	NR
2011	Ugu District [9]	KwaZulu-Natal	2 (NR)	<10%	NR	NR
2011	Ugu, iLembe, and Southern uThungulu (King Cetshwayo) districts [9]	KwaZulu-Natal	2 (NR)	≥50%	NR	NR
2011	Ugu, iLembe, and Southern uThungulu (King Cetshwayo) districts [9]	KwaZulu-Natal	47 (NR)	10–49%	NR	NR
2011	Ugu, iLembe, and Southern uThungulu (King Cetshwayo) districts [9]	KwaZulu-Natal	13 (NR)	<10%	NR	NR
2011–2013	iLlembe, Ugu and uThungulu (King Cetshwayo) districts [5]	KwaZulu-Natal	NR (1123)	26.0%	NR	
2012	Ugu District [28]	KwaZulu-Natal	3 (246) and 9 [d] (873)	20.4%	NR	14
2014	uMkhanyakude (Jozini Municipality) [29]	KwaZulu-Natal	10 (420)	40.2%	NR	NR
2015	uMkhanyakude District (Ndumo) [30]	KwaZulu-Natal	10 (320)	37.5%	NR	NR
2019	uMkhanyakude District [31]	KwaZulu-Natal	34 (1143) [e]	1.0%	0.9%	30.4
2004 2005 2005 2005 2005 2005	Vhembe District [32]	Limpopo	NR	>70% [f]	NR	NR
2005	Vhembe District [32]	Limpopo	1 (94)	36.2% [g]	NR	NR
2005	Vhembe District [32]	Limpopo	NR (148)	42% [h]	NR	NR
2005	Vhembe District [32]	Limpopo	1 (247)	86% [e]	NR	NR
2005	Vhembe District [32]	Limpopo	1 (191)	84% [e]	NR	NR
2005	Vhembe District [32]	Limpopo	1 (138)	78.2% [e]	NR	NR
2009–2013	Rob Ferreira Hospital Patients [33]	Mpumalanga	304 [i]	10.2%	NR	NR

NR = not reported. Population in rows 1,4, and 6–9 may partially overlap. [a] Public primary schools. [b] Measured during hot season. [c] Measured during cold season. [d] High schools. [e] Preschools and early childhood development (ECD) centres. [f] Retrospective laboratory data analysis for the year 2004 in major hospitals in Vhembe District (indicated as hospital A, B and C in the published paper) among patients who attended the main hospitals with urinary tract infection. [g] University of Venda students. [h] Primary school learners. [i] Appendix samples removed in theatre, Rob Ferreira Hospital for histological investigation (microscopy for schistosomiasis). Results were not classified according by schistosome species.

A school survey conducted during 2016–2019 by the South African Department of Health and University of KwaZulu-Natal found that schistosomiasis prevalence was moderate (between 10% and 50%) in 32 local municipalities within five provinces, excluding Gauteng (Johannesburg area) and North West provinces, whereas one local municipality and 8 schools were identified as high risk (more than 50% prevalence) in KwaZulu-Natal (Figure 1) [34–40].

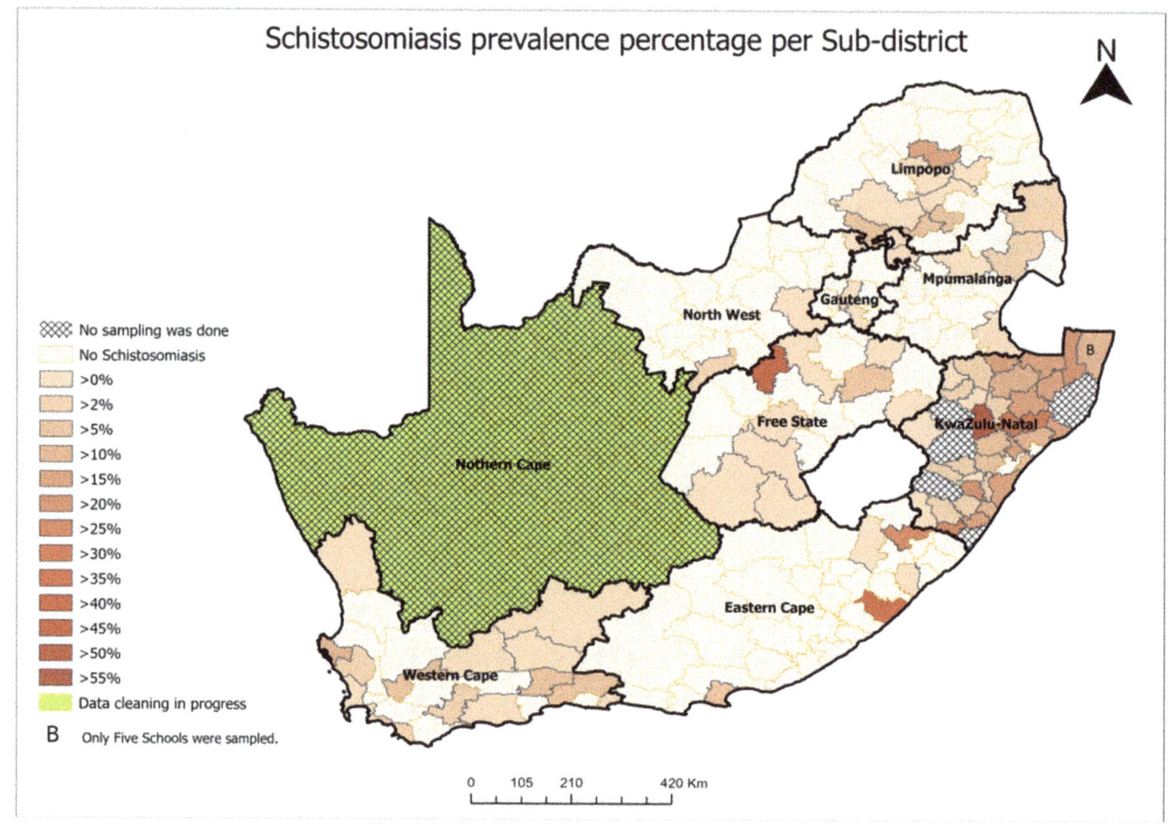

Figure 1. Distribution of schistosomiasis in South Africa.

Although findings from the mapping revealed low prevalence in most districts, it is important to note that the mapping exercise had some limitations related to sampling criteria given the focality of the disease, timing of specimen collection, and frequency of specimen collection, e.g., collecting a single sample only. These limitations result in difficulties in making inferences on the prevalence and intensity of infection, as well as the presence of female genital lesions and urinary tract disease [13,41].

As has been found for other urban areas, Gauteng province was found to be at low-risk for schistosomiasis and reports indicate that cases may have been imported from other provinces and countries [42–44]. FGS screening and management should therefore be performed even in provinces that do not have local schistosomiasis transmission [11]. Without schistosomiasis mass treatment interventions and screening measures, the FGS prevalence will increase surreptitiously and continue to have obvious negative impacts such as sub-fertility or infertility, ectopic pregnancy, spontaneous abortion, premature birth, low birth weight, maternal death, genital symptoms, and increased HIV susceptibility [3,4,45,46]. If mass treatment is implemented alongside individual case management, continuous provision of piped water and community awareness, a schistosomiasis elimination target of <1% proportion of heavy intensity infections with schistosomiasis can be achieved [45]. The purpose of this policy brief is to identify action points for the control of FGS in Africa.

2.3. Policy Actions and Implications

The South African Department of Health has not implemented a control programme despite its policy and implementation guidelines called "Regular treatment of school-going children for soil-transmitted helminth infections and Bilharzia" [47]. Similarly, many other countries have failed to follow through with 75% treatment coverage in endemic areas and women suffer the consequences (45). In South Africa, schistosomiasis management is limited to case-based treatment for those who seek medical care; however, genital schistosomal morbidity is seldom managed [9,47]. Some community members do not even seek medical care if they experience blood in the urine, because they believe this is normal [48]. Often, health professionals in South Africa report and take action upon discovering high-endemic sites of schistosomiasis, usually among school children, but there are no ongoing awareness programmes or systematic testing [49,50]. Moreover, the free generic treatment that is accredited by the WHO has been donated to all endemic countries but is not available to South Africans [51]. In an impassse, the South African government does not accept the WHO accreditation and the donating company does not think it should pay for the entire process of registering the donated medication in recipient countries [51].

In order to interrupt the life cycle of schistosome transmission and prevent the occurrence of new cases, many researchers recommend a combination of control measures (including a vector control strategy which has not been fully applied in most countries) [52–56]. This would positively contribute to controlling FGS. In brief, prevention of Female Genital Schistosomiasis and reduction of morbidity from this disease requires five main actions:

Action A: Schistosomiasis prevention by establishing a formal control programme and increasing access to treatment:

Preventive chemotherapy—Worldwide, there is currently no ongoing treatment programme for FGS [57,58]. Treatment with praziquantel targets the adult worm but has no effect on the ova in the tissues; therefore, treatment should be undergone 6–8 weeks after exposure, before massive ova deposition is caused FGS [8,58]. Mass drug administration (MDA) has been implemented in some countries against schistosomiasis and is recommended by the WHO to reduce morbidity and move towards elimination [45,58]. In addition, to prevent FGS, the following should be implemented:

- Preventive chemotherapy in girls against FGS alongside HPV vaccination (cervical cancer prevention) in schistosomiasis endemic schools [3,59,60]. This will secure at least one round of treatment to prevent FGS and promote awareness around the genital morbidity caused by the two diseases;
- Hot-spot targeted administration of praziquantel and morbidity screening [59,61,62]; and
- Use the opportunities to treat at the endemic primary health care facilities, provide regular treatment for adults, in addition to other community members at risk of infection and school children [59].

Administration of praziquantel in other health programmes—To improve access to treatment, it is recommended that praziquantel be administered to all individuals at risk of infection in the following health programmes:

- During promotion of sexual and reproductive health at the reproductive health clinics in endemic areas [15,57,59];
- Alongside antiretroviral therapy and pre-exposure prophylaxis for HIV/AIDS [3,15]; and
- Alongside cervical cancer screening.

Action B: FGS Screening–FGS screening and diagnosis should be accessible at the following various platforms in the health care system and the community:

FGS diagnosis—As a common differential diagnoses, it is recommended that all individuals with symptoms of STIs and cervical cancer be screened for FGS alongside cervical cancer screening and STI testing in all health care facilities of endemic countries; cases have been reported in South Africa, Zimbabwe, Mozambique, Cameroun, Ghana, Nigeria, Egypt, Kenya, Tanzania, and Madagascar, showing that the disease is likely present on the entire African continent [1,3,15,18,58,63–68].

- Index and secondary (surrounding) cases' management—FGS is treated with praziquantel and as individualized disease management. However, hot spot intervention for community members who use the same water source as the index case should be carried out [45].

- Establishment of sentinel sites—At present, only crude extrapolations of the burden of FGS are possible [3]. Therefore, sentinel sites should be established in all municipalities where schistosomiasis is endemic (Figure 1 for South Africa) to establish baseline prevalence of FGS and track progress toward elimination. At these sites, FGS screening and case management should be prioritized alongside cervical cancer screening. Regular monitoring and evaluation surveys every 3 or 5 years are also appropriate for the follow-up of the control progress.

Action C: Vector Control

Vector control strategy involves freshwater molluscicides (chemical control), physical removal of snails (the intermediate host for schistosomiasis), and environmental modification [45,52]. This requires identification of possible infection sites, and identification of intermediate host snails as described by Chris Appleton and Nelson Miranda [52]. The process involves the following:

- Regular testing of water bodies as schistosomiasis transmission sites through the collection of snails and identification of intermediate hosts of *S. haematobium* and *S. mansoni*, followed by examination through snail dissection, crushing, and analysis, e.g., by Polymerase Chain Reaction (PCR) [52];
- Chemical control methods (e.g., molluscicides to kill the snails) are applied in artificial water bodies, such as irrigation channels, ditches, and farm dams, but not in large dams, natural streams, rivers, and lakes [52];
- Environmental control involves the removal of vegetation to remove the snails' environment, depriving them of the sheltered niches they favour, and where appropriate, ensuring that the water flows faster than their tolerance limit of 0.3 m/s. Ideally, canals should be concrete lined and contoured, to encourage fast flow. Covering open canals near dwellings is a simple way of deterring human contact with the water [52]

Action D: Water, Sanitation, and Hygiene (WASH)–Provision of Continuous Piped Water

The provision of continuous piped water will reduce community members' exposure to contaminated fresh water during chores such as washing and collecting water for bathing and cooking. The provision of sanitation facilities will reduce miracidial contamination of water through urine, vaginal fluids, or stool. This strategy may a require massive cost outlay but it will be cost-beneficial in the long run; it requires a long-term goal for multi-sectoral collaboration [58,64]. The strategy involves:

- Identification of households that do not have access to proper sanitation and piped water;
- Installation and maintenance of proper sanitary structures and water pipes to these households;
- Household education on the effective use of sanitary structures and water to improve hygiene; and
- Continuous health promotion campaigns and awareness on the prevention of schistosomiasis and FGS, to influence both domestic and recreational behaviour change.

Action E: Creating Awareness of Schistosomiasis and FGS

The demand for medication may be low due to a lack of awareness and lack of referrals for treatment [69]. There has never been an awareness programme for FGS [20,25,70]. This cross-cutting activity will augment all the other actions. For effectiveness and yield of the desired outcome, an awareness programme must be implemented as a parallel programme prioritizing the following key areas:

- Training of those who use speculums in clinical practice (clinicians, gynaecologists, nurses, and midwives) in FGS diagnosis and treatment [22,45,64]. This should include awareness of their patient's status as an index case, active search for other cases, training in surveillance, and notification.

- Training of health promoters, risk communicators, environmental health practitioners, and communicable diseases control coordinators on schistosomiasis and FGS prevention and control;
- Training of community health workers (lay people) on schistosomiasis and FGS prevention and control;
- Training of water and sanitation officials in schistosomiasis and FGS prevention to assist with community awareness during installation and maintenance of sanitary structures and water pipes;
- Population awareness campaigns on schistosomiasis, FGS, and measures for prevention (including preventing exposure to risky freshwater) to influence behaviour change [45,66].

The relevant stakeholders for policy actions are summarized in Table 2, and are listed according to their interest and the power they have in the identified actions [71]:

- High power, highly interested stakeholders should be managed closely, be fully engaged, and it is important to make the greatest efforts to satisfy them;
- High power, less interested stakeholders are stakeholders that need to be kept satisfied, but not so much that they become bored with messages;
- Low power, highly interested stakeholders are stakeholders that should be kept informed with adequate information to ensure that no major issues are arising. Stakeholders in this category can often be very helpful with the details of the implementation; and
- Low power, less interested people are stakeholders that need to be monitored, but not bored with excessive communication.

As shown in Figure 2, the conceptual framework highlights various actions that can be implemented to decrease the burden of FGS, and reduce schistosomiasis and FGS prevalence. These actions are complementary, and can be implemented as a package.

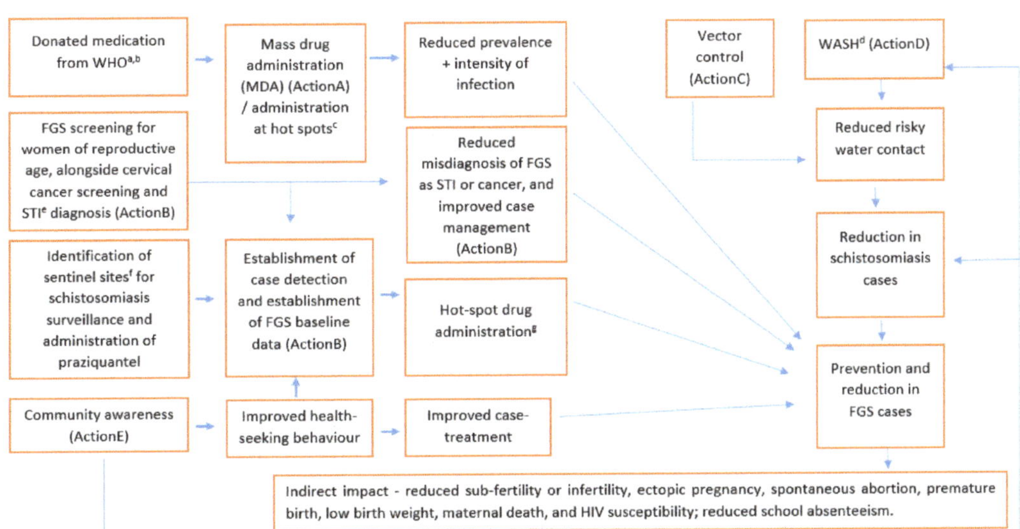

Figure 2. Conceptual framework to guide identification of actions for control of female genital schistosomiasis (FGS). [a] World Health Organization, [b] To our knowledge, approval is still required in South Africa to use the donated medication from WHO, [c] Places where clusters of cases are detected, [d] Water, sanitation and hygiene, [e] Sexually transmitted infection, [f] Health care facilities identified as sites for monitoring of disease pattern and progress of the control programme, [g] Administration of praziquantel at the emerging hot spots where clusters of cases are detected.

Table 2. Stakeholders for each policy action.

Policy Actions	Strategic Activities	Stakeholder at the Forefront with Both High Power and High Interest [a]	Other Stakeholders and Their Interest
Action A: FGS screening	• FGS diagnosis • Case management • Establishment of sentinel sites for surveillance	• Department of Health [a] • Gynaecological services, cervical cancer screening programme, STI clinics [a] • Department of Vector control and Public Health, National Institute for Communicable Diseases (NICD)	• Department of Education and School Governing Body have high interest [a] and low power • Community members including leaders, learners and their parents have high interest [a] and low power • Research community has high interest [a] and low power • The World Health Organization (WHO) has high interest and in-country low power
Action B: Treatment for Schistosomiasis	• Praziquantel (Mass drug administration/administration at hot spots/Case treatment)	• Department of Health [a] • Department of Education [a] • WHO	• School Governing Body [b] have high interest and low power • Community members including leaders, learners and their parents have high interest [a] and low power • South African Health Products Regulatory Authority have high power but low interest • Research community has high interest and low power
Action C: Vector Control	• Freshwater snail control with molluscicides (chemical control) • Physical removal of snails, and environmental modification	• Department of Health • Department of Environmental Affairs • Department of Vector control, NICD	• Department of Agriculture, Land Reform and Rural Development has high power and low interest • Research and Academic Institutions have high interest and less power • WHO has high interest and low power
Action D: Water, Sanitation and Hygiene (WASH)	• Provision of continuous piped water • Provision of proper sanitary facilities • Hygiene education • Training and awareness campaigns on schistosomiasis and FGS	• Department of Water and Sanitation • Department of Health • Municipalities and South African Local Government Authority	• Community members including leaders, learners and their parents have high interest [a] and low power • WHO has high interest and low power • Research and Academic Institutions have high interest and low power

Table 2. *Cont.*

Policy Actions	Strategic Activities	Stakeholder at the Forefront with Both High Power and High Interest [a]	Other Stakeholders and Their Interest
Action E: Creating Awareness of Schistosomiasis and FGS	• Training and awareness on schistosomiasis and FGS among health care professionals • Awareness campaigns in the community on schistosomiasis and FGS	• Department of Health • WHO • Department of Education • Department of Water and Sanitation • Municipalities and South African Local Government Authority • National Health Laboratory Service (control programme and diagnostics, research, profit and improved public health) • Research community and Academic Institution has interest in the control programme and improved public health • School Governing Body	• Department of Agriculture, Land Reform and Rural Development has less interest and low power • Department of Environmental Affairs has low interest and low power

Interest requires knowledge and awareness. Currently most health professionals do not know about FGS [11,70]. [a] When they are knowledgeable and aware of FGS, [b] Elected by parents.

3. Public Health Benefit

Health care worker knowledge, case-by-case therapy, and prevention chemotherapy with praziquantel have been proven to be effective and can improve individuals' health in the affected communities [45,72]. In 2016, schistosomiasis caused an estimated 24,000 deaths and 2.4 million disability-adjusted life years (DALYs) worldwide [45]. One study suggested that regular anti-schistosomal MDA with praziquantel could prevent genital schistosomiasis in more than 200,000 young women in rural KwaZulu-Natal province schools in South Africa, and that by treating and preventing FGS, it would be possible to prevent more than 5000 HIV infections in adolescent girls and young women [9].

The effects would be difficult to measure as many components influence behaviour and HIV transmission, and none can be put on hold. The expected outcomes of the programme are reduced number of FGS and schistosomiasis cases, as well as reduced school absenteeism. Therefore, indirect indicators that can be used to measure public health benefit from the recommended control actions could be number of diagnosed FGS and schistosomiasis cases or unconfirmed STI symptom cases, as well as the rate of school absenteeism. "Doctor shopping" surveys for genital symptoms could be measured. A study conducted in Bandanyenje Primary School in the Manicaland Province in Zimbabwe found that annual praziquantel treatment delivered to school children over 2 years of age had a significant impact on the reduction of prevalence, intensity of infection, and reinfection of *S. haematobium* infection [73]. In another study, women who had received anti-schistosomal treatment in childhood had much less genital morbidity as adults [8]. However, such long-term effects would be difficult to measure unless FGS becomes a notifiable disease.

It is not clear whether the availability of continuous piped water will completely influence behaviour change and community members will stop exposing themselves to risky fresh water, for example, even children with taps continue to use rivers and dams for recreational purpose and laundry might be easier to complete in a river, than in small basins at home [7]. This might require vigorous community awareness and health promotion campaigns.

4. Ethics/Equity

The Department of Health South Africa has surveyed schistosomiasis and identified the communities at risk of infection; lack of schistosomiasis control services in these communities is unethical [34–37]. FGS has also been shown to have a significant detrimental influence on the HIV pandemic among untreated adolescent girls and young women, including those with historical unsafe water contact and who move to urban areas [9,74]. The majority of communities at risk of infection also have difficulty obtaining health services due to transportation issues or a lack of awareness about disease prevention. As a result, in order to ensure equitable distribution of resources, these populations must be prioritized for preventive chemotherapy, case management, awareness, and reproductive health care. Furthermore, the Department of Water and Sanitation is a required stakeholder for the Department of Health to campaign for piped water.

5. Administrative Feasibility and Budgetary Feasibility

FGS should be investigated where visual inspections are carried out using a speculum and a good light source. As a disease affecting women of reproductive age, FGS should therefore fall under the jurisdiction of cervical cancer screening, reproductive health, primary health care, and HIV prevention programmes [15]. The schistosomiasis prevention programme is within the mandates of the communicable diseases control units of the national Departments of Health, including development and implementation of control strategies. However, currently, generic medication for preventive chemotherapy is donated by Merck, via the World Health Organization, to all the affected countries. South Africa is yet to have the donated medication registered in the country. As a temporary measure, one can apply to use donated praziquantel as an exception (under Section 21 of the South African Health Product Regulatory Authority) [75]. Furthermore, South Africa has a tender for the (non-generic) schistosomiasis medication that is on the essential drug list; this is more expensive [76]. However, three pharmaceutical companies have the WHO prequalification for praziquantel (Macleods, Medopharm, and Hetero Labs) opening for competitive prices for procurement by the country [77]. HPV vaccination is carried out by the Department of Basic Education and they have already included annual school-based deworming for soil-transmitted helminths (STHs) in partnership with the Department of Health; schistosomiasis deworming could be included into this existing programme utilizing the existing resources. Stand-alone deworming programmes are expensive, and the cost per treated person is very high if many refuse [78]. Integrating FGS diagnosis and treatment into other programmes will be more cost-efficient, making it more fundable, and time-efficient.

6. Political Feasibility of Donated Medication in South Africa

The South African Health Product Regulatory Authority requires that all health products be approved and registered before entry and use. On the other hand, South Africa and many other countries are Member States of the World Health Assembly, and Ministers of Health have endorsed the regulations relating to the control of schistosomiasis with emphasis on mass drug administration. Ministers of Health should therefore support advocacy efforts for improved access to treatment, including entry into South Africa. In the past, TAC launched a successful campaign and put pressure on the Government of South Africa to make available AIDS treatment in public facilities [79]. TAC's vision continues to be "engaging in monitoring, advocacy, and campaigning within the healthcare system to ensure that all public healthcare users can access quality and dignified healthcare—and that all people with HIV and TB can access prevention, treatment, care, and support services" [79]. Schistosomiasis is associated with increased risk for HIV transmission, and organizations such as Treatment Action Campaign (TAC) should be targeted for advocacy [16,63,80–84].

Table 3 shows the weighing score for each policy action [85]; the actions are in line with the WHO schistosomiasis control and elimination recommendations and experiences from countries such as China [45,60,86–91]. All actions were weighted as low, medium, or high for each criterion mentioned above, and based on the requirement for shorter or longer term activities for each action to impact on the expected outcome of the programme (reduced or elimination of FGS and schistosomiasis).

Table 3. Policy actions matrix for Africa.

Action	Criteria			
	Public Health Impact (Efficacy)	**Ethics/Equity**	**Administrative Feasibility**	**Budgetary Feasibility**
A. Prevention and increased access to treatment	High	High	High	High
B. FGS diagnosis	High	High	High	High
C. Vector control	High	Medium	Medium	Medium
D. Water, sanitation and hygiene (WASH)	High	High	Low	Low
E. Creating awareness	High	Medium	High	High

Low weight means poor, medium means moderate, and high weight means good.

7. Implementation Challenges and Recommendations

7.1. Adaptive Challenges and Recommendations

The WHO-accredited Tablet Pole (measuring height) replaces the weight scale in most mass treatment campaigns [92]. In South Africa, it has been found to be inaccurate in determining the praziquantel dose due to the presence of overweight/obese children [93]. Precision can, however, be improved by adding an extra tablet of praziquantel to the standard dose.

In most countries, medication is dispensed by teachers as recommended by the WHO [92]. In South Africa, however, praziquantel is a "schedule 4" drug and must be dispensed by nurses or doctors. Teachers are currently dispensing mebendazole for soil-transmitted helminths. However, they have not taken on anti-schistosomal medication, because it is "illegal" and furthermore, it could be seen as a significant burden and additional obligation for teachers to use a weight scale or the WHO Tablet Pole if not properly explained [93]. In South Africa, therefore, school health nurses dispense praziquantel, greatly increasing the cost of FGS prevention [9,78,94]. Regulatory issues must be addressed and subsequently, teachers, their unions, and school governing bodies should be addressed [92].

Finally, girls could, alongside HPV vaccination/deworming to prevent cervical cancer/STHs, also be given praziquantel, to prevent FGS.

7.2. Regulatory Challenges for Praziquantel, Specifically for South Africa and Recommendations

In South Africa, praziquantel is dispensed by registered nurses or doctors even though it is very safe and has been dispensed by teachers for four decades in Asia, South America, and the rest of Africa. The South African regulation drives the cost up and highly impractical in mass treatment scenarios. Furthermore, studies have found that there is better compliance if teachers are dispensing it [92]. Additionally, two brands of praziquantel are registered in South Africa but not the WHO-donated drug. As a result, the cost of mass treatment is much higher in South Africa than in similar countries. Therefore, the following is recommended:

- Praziquantel should be down-regulated from schedule 4 to schedule 1.
- The Department of Health must apply to the South African Health Products Regulatory Authority under Section 21 of the South African Health Product Regulatory Authority for the admission of donated pharmaceuticals.
- The disease is neglected and will require dedicated coordinators. Staff turnover may have an influence on program implementation. Therefore, integration into the existing programmes and improving awareness among health care workers are essential.
- FGS should be incorporated in the "Regulations relating to the surveillance and the control of notifiable medical conditions" [12].
- Registration of the WHO donated praziquantel, Cesol®, Merck, Mexico, should be prioritized instead of using Section 21 of the South African Health Product Regulatory Authority exemption on an annual basis. This can be achieved through strong advocacy and evidence-based motivation.
- Ensure continued availability of praziquantel in all known endemic primary health care facilities, mother and child clinics, cervical cancer screening sites, HPV vaccination programmes, PrEP, and other HIV prevention programmes.

8. Technical Issues and Recommendations

Colposcopes are generally not available in low-resource settings for visualization of the lesions, and the training of clinicians and other health care workers on schistosomiasis and FGS diagnostic skills should be carried out [17,20]. Health information officers, data collectors, and personnel must be trained in order to increase knowledge and reporting/notification and data collection tool must be developed.

9. Laboratory Analysis for FGS

Clinical (visual) recognition is essential for FGS diagnosis and cannot be replaced by urine microscopy. However, PCR of genital specimens may support the FGS diagnosis. The Circulating Anodic Antigen (CAA) test (Leiden University Medical Centre, The Netherland and Ampath, South Africa) may be used to monitor progress towards elimination at the sentinel sites and may be useful for screening as an alternative to urinary microscopy, especially where prevalence and intensity of infection are low. The CAA has been found to be a highly sensitive and 100% specific test, capable to detect extremely low concentrations of the parasite-excreted CAA, potentially down to the level of a single worm infection [95,96].

Actions A (FSG diagnosis), B (improve access to treatment), and D (water, sanitation and hygiene) are recommended as immediate priorities since they will lead to the decrease in FGS burden, as well as the decrease or elimination in schistosomiasis and FGS. Actions C (vector control) and E (creating awareness) are recommended as long-term measures to support sustainability of the control programme and eventual elimination.

10. Conclusions

Africa lacks an FGS control programme and there is lack of knowledge among health care workers and communities. Integration of FGS diagnosis into the existing health care system presents opportunities to aid effective and efficient implementation of the programme [10,11].

Clinical misdiagnosis will result in untreated FGS in women and girls with symptoms of unknown cause [69]. FGS patients should be viewed as index cases with associated infections. It has been shown that local transmission may occur in poor urban settings and it is important to determine the source of urban infections and manage them [42,97–102].

In order to move towards elimination, a 2030 WHO goal, treatment programmes, and mass drug administration should be carried out at schools, in reproductive health facilities, and in the impacted communities [45,58]. To increase access to treatment in South Africa through the acceptance of WHO accredited donated medicine, resolution is likely to require a political stance or a court case. Improved access to schistosomiasis treatment will decrease the risk of new FGS and enhance the health in the affected communities. To realize the wish for treatment, community awareness must run concurrently with training of health care professionals as well as community health care workers. Although vector control, water, sanitation, and hygiene initiatives are expensive and will only be effectuated in the long term, they are vital; community awareness will play a large role by offering knowledge on actions for reducing unsafe water exposure.

Author Contributions: Writing—original draft preparation, T.G.N.; data curation, T.G.N.; writing—review and editing, T.G.N., T.E.F., M.K.G., A.G.D., S.N., and E.F.K.; supervision, S.N. and E.F.K.; project administration, E.F.K. and S.N. All authors have read and agreed to the published version of the manuscript.

Funding: The research leading to these results has received co-funding from the European Research Council under the European Union's Horizon 2020/ERC Grant agreement no. 101057853 (DUALSAVE-FGS).

Data Availability Statement: Data sharing is not applicable; no new data created.

Acknowledgments: We are appreciative the technical support from Alison C. Cullen from the International Program in Public Health Leadership, University of Washington. Dorman Chimhamhiwa and Rilwele Rabambi from Right To Care is thanked for generating the map.

Conflicts of Interest: The authors declare that there is no conflict of interest.

References

1. Kjetland, E.F.; Gwanzura, F.; Ndhlovu, P.D.; Mduluza, T.; Gomo, E.; Mason, P.R.; Midzi, N.; Friis, H.; Gundersen, S.G. Simple clinical manifestations of genital Schistosoma haematobium infection in rural Zimbabwean women. *Am. J. Trop. Med. Hyg.* **2005**, *72*, 311–319. [CrossRef]
2. Jordens, M. The Global Burden of Female Genital Schistosomiasis. *infoNTD*. 2019. Available online: https://www.infontd.org/resource/global-burden-female-genital-schistosomiasis (accessed on 17 September 2022).
3. Christinet, V.; Lazdins-Helds, J.K.; Stothard, J.R.; Reinhard-Rupp, J. Female genital schistosomiasis (FGS): From case reports to a call for concerted action against this neglected gynaecological disease. *Int. J. Parasitol.* **2016**, *46*, 395–404. [CrossRef]
4. Freer, J.B.; Bourke, C.D.; Durhuus, G.H.; Kjetland, E.F.; Prendergast, A.J. Schistosomiasis in the first 1000 days. *Lancet Infect. Dis.* **2018**, *18*, e193–e203. [CrossRef]
5. Galappaththi-Arachchige, H.N.; Holmen, S.; Koukounari, A.; Kleppa, E.; Pillay, P.; Sebitloane, M.; Ndhlovu, P.; Van Lieshout, L.; Vennervald, B.J.; Gundersen, S.G.; et al. Evaluating diagnostic indicators of urogenital Schistosoma haematobium infection in young women: A cross sectional study in rural South Africa. *PLoS ONE* **2018**, *13*, e0191459. [CrossRef]

6. van Bogaert, L.J. Schistosomiasis—An endemic but neglected tropical disease in Limpopo. *S. Afr. Med. J.* **2010**, *100*, 788–789. [CrossRef]
7. Hegertun, I.E.A.; Sulheim Gundersen, K.M.; Kleppa, E.; Zulu, S.G.; Gundersen, S.G.; Taylor, M.; Kvalsvig, J.D.; Kjetland, E.F. S. haematobium as a Common Cause of Genital Morbidity in Girls: A Cross-sectional Study of Children in South Africa. *PLoS Negl. Trop. Dis.* **2013**, *7*, e2104. [CrossRef]
8. Kjetland, E.F.; Ndhlovu, P.D.; Kurewa, E.N.; Midzi, N.; Gomo, E.; Mduluza, T.; Friis, H.; Gundersen, S.G. Prevention of gynecologic contact bleeding and genital sandy patches by childhood anti-schistosomal treatment. *Am. J. Trop. Med. Hyg.* **2008**, *79*, 79–83. [CrossRef]
9. Livingston, M.; Pillay, P.; Zulu, S.G.; Sandvik, L.; Kvalsvig, J.D.; Gagai, S.; Galappaththi-Arachchige, H.; Kleppa, E.; Ndhlovu, P.D.; Vennervald, B.J.; et al. Mapping Schistosoma haematobium for novel interventions against Female Genital Schistosomiasis and associated HIV risk in KwaZulu-Natal, South Africa. *Am. J. Trop. Med. Hyg.* **2021**, *104*, 2055–2064. [CrossRef]
10. Gyapong, J.O.; Gyapong, M.; Yellu, N.; Anakwah, K.; Amofah, G.; Bockarie, M.; Adjei, S. Integration of control of neglected tropical diseases into health-care systems: Challenges and opportunities. *Lancet* **2010**, *375*, 160–165. [CrossRef]
11. Engels, D.; Hotez, P.J.; Ducker, C.; Gyapong, M.; Bustinduy, A.L.; Secor, W.E.; Harrison, W.; Theobald, S.; Thomson, R.; Gamba, V.; et al. Integration of prevention and control measures for female genital schistosomiasis, HIV and cervical cancer. *Bull. World Health Organ.* **2020**, *98*, 615–624. [CrossRef]
12. Department of Health. *Regulations Relating to the Surveillance and the Control of Notifiable Medical Conditions*; Government Gazette: Pretoria, South Africa, 2017; Volume 630, pp. 1–32.
13. Christensen, E.E.; Taylor, M.; Zulu, S.G.; Lillebo, K.; Gundersen, S.G.; Holmen, S.; Kleppa, E.; Vennervald, B.J.; Ndhlovu, P.D.; Kjetland, E.F. Seasonal variations in schistosoma haematobium egg excretion in school-age girls in rural Kwazulu-Natal province, South Africa. *S. Afr. Med. J.* **2018**, *108*, 352–355. [CrossRef]
14. Galappaththi-Arachchige, H.N.; Zulu, S.G.; Kleppa, E.; Lillebo, K.; Qvigstad, E.; Ndhlovu, P.; Vennervald, B.J.; Gundersen, S.G.; Kjetland, E.F.; Taylor, M. Reproductive health problems in rural South African young women: Risk behaviour and risk factors. *Reprod. Health* **2018**, *15*, 1–10. [CrossRef]
15. UNAIDS; World Health Organization. No More Neglect. Female Genital Schistosomiasis and HIV. In *Integrating Reproductive Health Interventions to Improve Women's Lives*; WHO: Geneva, Switzerland, 2019. Available online: https://www.unaids.org/sites/default/files/media_asset/female_genital_schistosomiasis_and_hiv_en.pdf (accessed on 12 March 2020).
16. Magaisa, K.; Taylor, M.; Kjetland, E.F.; Naidoo, P.J. A review of the control of schistosomiasis in South Africa. *S. Afr. J. Sci.* **2015**, *111*, 1–6. [CrossRef]
17. Swai, B.; Poggensee, G.; Mtweve, S.; Krantz, I. Female genital schistosomiasis as an evidence of a neglected cause for reproductive ill-health: A retrospective histopathological study from Tanzania. *BMC Infect. Dis.* **2006**, *23*, 134. [CrossRef]
18. Toller, A.; Scopin, A.C.; Apfel, V.; Prigenzi, K.C.; Tso, F.K.; Focchi, G.R.; Speck, N.; Ribalta, J. An interesting finding in the uterine cervix: Schistosoma hematobium calcified eggs. *Autops. Case Rep.* **2015**, *5*, 41–44. [CrossRef]
19. Mazigo, H.D.; Samson, A.; Lambert, V.J.; Kosia, A.L.; Ngoma, D.D.; Murphy, R.; Matungwa, D.J. "We know about schistosomiasis but we know nothing about FGS": A qualitative assessment of knowledge gaps about female genital schistosomiasis among communities living in Schistosoma haematobium endemic districts of Zanzibar and Northwestern Tanzania. *PLoS Negl. Trop. Dis.* **2021**, *15*, e0009789. [CrossRef]
20. Søfteland, S.; Sebitloane, M.H.; Taylor, M.; Roald, B.B.; Holmen, S.; Galappaththi-Arachchige, H.N.; Gundersen, S.G.; Kjetland, E.F. A systematic review of handheld tools in lieu of colposcopy for cervical neoplasia and female genital schistosomiasis. *Int. J. Gynecol. Obstet.* **2021**, *153*, 190–199. [CrossRef]
21. Norseth, H.M.; Ndhlovu, P.D.; Kleppa, E.; Randrianasolo, B.S.; Jourdan, P.M.; Roald, B.; Holmen, S.D.; Gundersen, S.G.; Bagratee, J.; Onsrud, M.; et al. The colposcopic atlas of schistosomiasis in the lower female genital tract based on studies in Malawi, Zimbabwe, Madagascar and South Africa. *PLoS Negl. Trop. Dis.* **2014**, *8*, e3229. [CrossRef]
22. Mbabazi, P.S.; Vwalika, B.; Randrianasolo, B.S.; Roald, B.; Ledzinski, D.; Olowookorun, F.; Hyera, F.; Galappaththi-Arachchige, H.N.; Sebitloane, M.; Taylor, M.; et al. World Health Organisation Female Genital Schistosomiasis. In *A Pocket Atlas for Clinical Health-care Professionals*; WHO: Geneva, Switzerland, 2015; Volume 2015, ISBN 978 92 4 150929 9. Available online: http://apps.who.int/iris/bitstream/10665/180863/1/9789241509299_eng.pdf (accessed on 20 May 2022).
23. Kjetland, E.F.; Kurewa, E.N.; Ndhlovu, P.D.; Midzi, N.; Gwanzura, L.; Mason, P.R.; Gomo, E.; Sandvik, L.; Mduluza, T.; Friis, H.; et al. Female genital schistosomiasis—A differential diagnosis to sexually transmitted disease: Genital itch and vaginal discharge as indicators of genital S. haematobium morbidity in a cross-sectional study in endemic rural Zimbabwe. *Trop. Med. Int. Health* **2008**, *13*, 1509–1517. [CrossRef]
24. Kjetland, E.F.; Ndhlovu, P.D.; Mduluza, T.; Deschoolmeester, V.; Midzi, N.; Gomo, E.; Gwanzura, L.; Mason, P.R.; Vermorken, J.B.; Friis, H.; et al. The effects of genital Schistosoma haematobium on human papillomavirus and the development of neoplasia after 5 years in a Zimbabwean population. A pilot study. *Eur. J. Gynec. Oncol.* **2010**, *31*, 169–173.
25. Kjetland, E.F.; Hegertun, I.E.A.; Baay, M.F.D.; Onsrud, M.; Ndhlovu, P.D.; Taylor, M. Genital schistosomiasis and its unacknowledged role on HIV transmission in the STD intervention studies. *Int. J. STD AIDS* **2014**, *25*, 705–715. [CrossRef] [PubMed]

26. Molvik, M.; Helland, E.; Zulu, S.G.; Kleppa, E.; Lillebo, K.; Gundersen, S.G.; Kvalsvig, J.D.; Taylor, M.; Kjetland, E.F.; Vennervald, B.J. Co-infection with Schistosoma haematobium and soil-transmitted helminths in rural South Africa. *S. Afr. J. Sci.* **2017**, *113*, 1–6. [CrossRef]

27. Zulu, S.G.; Kjetland, E.F.; Gundersen, S.G.; Taylor, M. Prevalence and intensity of neglected tropical diseases (schistosomiasis and soil-transmitted helminths) amongst rural female pupils in Ugu district, KwaZulu-Natal, South Africa. *S. Afr. J. Infect. Dis.* **2020**, *35*, 1–7. [CrossRef]

28. Kildemoes, A.O.; Kjetland, E.F.; Zulu, S.G.; Taylor, M.; Vennervald, B.J. Schistosoma haematobium infection and asymptomatic bacteriuria in young South African females. *Acta Trop.* **2015**, *144*, 19–23. [CrossRef] [PubMed]

29. Rubaba, O.; Chimbari, M.J.; Soko, W.; Manyangadze, T.; Mukaratirwa, S. Validation of a urine circulating cathodic antigen cassette test for detection of Schistosoma haematobiumin uMkhanyakude district of South Africa. *Acta Trop.* **2018**, *182*, 161–165. [CrossRef]

30. Kabuyaya, M.; Chimbari, M.J.; Manyangadze, T.; Mukaratirwa, S. Schistosomiasis risk factors based on the infection status among school-going children in the Ndumo area, uMkhanyakude district, South Africa. *S. Afr. J. Infect. Dis.* **2017**, *32*, 67–72. [CrossRef]

31. Sacolo-Gwebu, H.; Chimbari, M.; Kalinda, C. Prevalence and risk factors of schistosomiasis and soil-transmitted helminthiases among preschool aged children (1–5 years) in rural KwaZulu-Natal, South Africa: A cross-sectional study. *Infect. Dis. Poverty* **2019**, *8*, 47. [CrossRef]

32. Samie, A.; Nchachi, D.J.; Obi, C.L.; Igumbor, E.O. Prevalence and temporal distribution of Schistosoma haematobium infections in the Vhembe district, Limpopo Province, South Africa. *Afr. J. Biotechnol.* **2010**, *9*, 7157–7164. [CrossRef]

33. Botes, S.N.; Ibirogba, S.B.; McCallum, A.D.; Kahn, D. Schistosoma prevalence in appendicitis. *World J. Surg.* **2015**, *39*, 1080–1083. [CrossRef]

34. National Department of Health. *Mpumalanga NTD School Prevalence*; National Department of Health: Pretoria, South Africa, 2018.

35. National Department of Health. *KwaZulu-Natal High Risk NTD School Prevalence*; National Department of Health: Pretoria, South Africa, 2016.

36. National Department of Health. *Eastern Cape NTD School Prevalencel*; National Department of Health: Pretoria, South Africa, 2017.

37. National Department of Health. *Limpopo NTD School Prevalence*; National Department of Health: Pretoria, South Africa, 2016.

38. National Department of Health. *KwaZulu-Natal NTD School Prevalence*; National Department of Health: Pretoria, South Africa, 2016.

39. National Department of Health. *Schistosomiasis and Soil-Transmitted Helminthiasis Mapping in Gauteng Province*; National Department of Health: Pretoria, South Africa, 2019.

40. National Department of Health. *Schistosomiasis and Soil-Transmitted Helminthiasis Mapping in North West Province*; National Department of Health: Pretoria, South Africa, 2019.

41. Wolmarans, C.; De Kock, K. Eier-ekskresiepatrone as kriterium vir die bepaling van prevalensie en intensiteit van besmetting by manlike en vroulike individue in verskillende ouderdomsgroepe in n skistosoomendemiese gebied in die Limpopoprovinsie. *Suid-Afrik. Tydskr. Vir Nat. En Tegnol.* **2010**, *29*, 130–144. [CrossRef]

42. Firmo, J.O.; Lima Costa, M.F.; Guerra, H.L.; Rocha, R.S. Urban schistosomiasis: Morbidity, sociodemographic characteristics and water contact patterns predictive of infection. *Int. J. Epidemiol.* **1996**, *25*, 1292–1300. [CrossRef] [PubMed]

43. Kloos, H.; Correa-Oliveira, R.; Oliveira Quites, H.F.; Caetano Souza, M.C.; Gazzinelli, A. Socioeconomic studies of schistosomiasis in Brazil: A review. *Acta Trop.* **2008**, *108*, 194–201. [CrossRef] [PubMed]

44. Ndamba, J.; Chidimu, M.G.; Zimba, M.; Gomo, E.; Munjoma, M. An investigation of the schistosomiasis transmission status in Harare. *Cent. Afr. J. Med.* **1994**, *40*, 337–342. [PubMed]

45. World Health Organisation. *Ending the Neglect to Attain the Sustainable Development Goals: A Road Map for Neglected Tropical Diseases 2021–2030*; World Health Organisation: Geneva, Switzerland, 2020; ISBN 9789240010352.

46. Kjetland, E.F.; Leutscher, P.D.C.; Ndhlovu, P.D. A review of female genital schistosomiasis. *Trends Parasitol.* **2012**, *28*, 58–65. [CrossRef] [PubMed]

47. National Department of Health. Regular Treatment of School-Going Children for Soil-Transmitted Helminth Infections and Bilharzia. In *Policy and Implementation Guidelines*; The South African National Department of Health: Pretoria, South Africa, 2008; pp. 1–61.

48. Feldmeier, H.; Poggensee, G.; Krantz, I. A synoptic inventory of needs for research on women and tropical parasitic diseases. II. Gender-related biases in the diagnosis and morbidity assessment of schistosomiasis in women. *Acta Trop.* **1993**, *55*, 139–169. [CrossRef]

49. KwaZulu-Natal Department of Health. *Bilharzia Outbreak in KwaZulu-Natal Ethekwini Municipality*; National Department of Health: Pretoria, South Africa, 2021.

50. KwaZulu-Natal Department of Health. *Bilharzia Outbreak in KwaZulu-Natal Zululand District*; National Department of Health: Pretoria, South Africa, 2017.

51. Berge, S.T.; Kabatereine, N.B.; Gundersen, S.G.; Taylor, M.; Kvalsvig, J.D.; Mkhize-Kwitshana, Z.; Jinabhai, C.C.; Kjetland, E.F. Generic praziquantel in South Africa: The necessity for policy change to avail cheap, safe and efficacious schistosomiasis drugs to the poor, rural population. *S. Afr J. Epidemiol. Infect.* **2011**, *26*, 22–25. [CrossRef]

52. Appleton, C.; Miranda, N. Locating bilharzia transmission sites in South Africa: Guidelines for public health personnel. *South. Afr. J. Infect. Dis.* **2015**, *30*, 95–102. [CrossRef]

53. Fenwick, A. Host-parasite relations and implications for control. *Adv. Parasitol.* **2009**, *68*, 247–261.

54. Dabo, A.; Doucoure, B.; Koita, O.; Diallo, M.; Kouriba, B.; Klinkert, M.Q.; Doumbia, S.; Doumbo, O. Reinfection with Schistosoma haematobium and mansoni despite repeated praziquantel office treatment in Niger, Mali. *Med. Trop.* **2000**, *60*, 351–355.

55. Davis, A. The Professor Gerald Webbe Memorial Lecture: Global control of schistosomiasis. *Trans. R. Soc. Trop. Med. Hyg.* **2000**, *94*, 609–615. [CrossRef]

56. Clark, T.E.; Appleton, C.C.; Kvalsvig, J.D. Schistosomiasis and the use of indigenous plant molluscicides: A rural South African perspective. *Acta Trop.* **1997**, *66*, 93–107. [CrossRef]

57. Bengu, M.D.; Dorsamy, V.; Moodley, J. Schistosomiasis infections in South African pregnant women: A review. *S. Afr. J. Infect. Dis.* **2020**, *35*, 7. [CrossRef]

58. van Bogaert, L. Case of invasive adenocarcinoma of the cervix in a human immunodeficiency virus and schistosome co-infected patient. *S. Afr. J. Gynaecol. Oncol.* **2014**, *6*, 5–6. [CrossRef]

59. World Health Organisation. *Deworming Adolescent Girls and Women of Reproductive Age: Policy Brief*; World Health Organisation: Geneva, Switzerland, 2021.

60. World Health Organization. *WHO Guideline on Control and Elimination of Human Schistosomiasis*; World Health Organisation: Geneva, Switzerland, 2022; p. 142. ISBN 978 92 4 004160 8. Available online: https://www.who.int/publications/i/item/9789240041608?msclkid=66e125e4aa9911eca97802b2699b0852 (accessed on 19 June 2022).

61. Cribb, D.M.; Clarke, N.E.; Doi, S.A.R.; Nery, S.V. Differential impact of mass and targeted praziquantel delivery on schistosomiasis control in school-aged children: A systematic review and meta-analysis. *PLoS Negl. Trop. Dis.* **2019**, *13*, e0007808. [CrossRef]

62. Mawa, P.A.; Kincaid-Smith, J.; Tukahebwa, E.M.; Webster, J.P.; Wilson, S. Schistosomiasis Morbidity Hotspots: Roles of the Human Host, the Parasite and Their Interface in the Development of Severe Morbidity. *Front. Immunol.* **2021**, *12*, 635869. [CrossRef]

63. Brodish, P.H.; Singh, K. Association between Schistosoma haematobium exposure and Human Immunodeficiency Virus infection among females in Mozambique. *Am. J. Trop. Med. Hyg.* **2016**, *94*, 1040–1044. [CrossRef]

64. Pinto, S.Z.; Friedman, R.; Van Den Berg, E.J. A case of paediatric bladder bilharzioma in Johannesburg, South Africa. *Clin. Case Rep.* **2019**, *7*, 1890–1894. [CrossRef]

65. van Bogaert, L.-J.J. Biopsy-diagnosed female genital schistosomiasis in rural Limpopo, South Africa. *Int. J. Gynaecol. Obstet.* **2011**, *115*, 75–76. [CrossRef]

66. Yirenya-Tawiah, D.; Amoah, C.; Apea-Kubi, K.A.; Dade, M.; Ackumey, M.; Annang, T.; Mensah, D.Y.; Bosompem, K.M. A survey of female genital schistosomiasis of the lower reproductive tract in the volta basin of ghana. *Ghana Med. J.* **2011**, *45*, 16–21. [CrossRef]

67. Randrianasolo, B.S.; Jourdan, P.M.; Ravoniarimbinina, P.; Ramarokoto, C.E.; Rakotomanana, F.; Ravaoalimalala, V.E.; Gundersen, S.G.; Feldmeier, H.; Vennervald, B.J.; Van Lieshout, L.; et al. Gynecological manifestations, histopathological findings, and schistosoma-specific polymerase chain reaction results among women with schistosoma haematobium infection: A cross-sectional study in Madagascar. *J. Infect. Dis.* **2015**, *212*, 275–284. [CrossRef]

68. Ekpo, U.F.; Odeyemi, O.M.; Sam-Wobo, S.O.; Onunkwor, O.B.; Mogaji, H.O.; Oluwole, A.S.; Abdussalam, H.O.; Stothard, J.R. Female genital schistosomiasis (FGS) in Ogun State, Nigeria: A pilot survey on genital symptoms and clinical findings. *Parasitol. Open* **2017**, *3*, e10. [CrossRef]

69. Jacobson, J.; Pantelias, A.; Williamson, M.; Kjetland, E.F.; Krentel, A.; Gyapong, M.; Mbabazi, P.S.; Djirmay, A.G. Addressing a silent and neglected scourge in sexual and reproductive health in Sub-Saharan Africa by development of training competencies to improve prevention, diagnosis, and treatment of female genital schistosomiasis (FGS) for health workers. *Reprod. Health* **2022**, *19*, 20. [CrossRef] [PubMed]

70. Gyapong, M.; Marfo, B.; Theobold, S.; Hawking, K.; Page, S.; Osei-Atweneboanaa, M.; Stothard, J. The double jeopardy of having female genital schistosomiasis in poorly informed and resourced health systems in Ghana. *Trop. Med. Int. Health* **2015**, *20*, 7.

71. Mind Tools Content Team. Stakeholder Analysis. Available online: https://www.mindtools.com/pages/article/newPPM_07.htm (accessed on 21 August 2021).

72. Turner, H.C.; French, M.D.; Montresor, A.; King, C.H.; Rollinson, D.; Toor, J. Economic evaluations of human schistosomiasis interventions: A systematic review and identification of associated research needs. *Wellcome Open Res.* **2020**, *5*, 45. [CrossRef] [PubMed]

73. Chisango, T.J.; Ndlovu, B.; Vengesai, A.; Nhidza, A.F.; Sibanda, E.P.; Zhou, D.; Mutapi, F.; Mduluza, T. Benefits of annual chemotherapeutic control of schistosomiasis on the development of protective immunity. *BMC Infect. Dis.* **2019**, *19*, 219. [CrossRef] [PubMed]

74. Hotez, P.J.; Fenwick, A.; Kjetland, E.F. Africa's 32 Cents Solution for HIV/AIDS. *PLoS Negl. Trop. Dis.* **2009**, *3*, e430. [CrossRef] [PubMed]

75. SAHPRA. *2. 52 Section 21 Access to Unregistered Medicines*; South African Health Products Regulatory Authority: Pretoria, South Africa, 2020; Volume 1965, pp. 1–15.

76. National Department of Health. *Standard Treatment Guidelines and Essential Medicines List for South Africa: Primary Health Care Level*, 7th ed.; The South African National Department of Health: Pretoria, South Africa, 2020; pp. 1–561.

77. World Health Organization. Prequalification of Medical Products (IVDs, Medicines, Vaccines and Immunization Devices, Vector Control). Available online: https://extranet.who.int/pqweb/medicines/prequalified-lists/finished-pharmaceutical-products?label=&field_medicine_applicant=&field_medicine_fpp_site_value=&search_api_aggregation_1=&field_medicine_pq_date%5Bdate%5D=&field_medicine_pq_date_1%5Bdate%5D=&fi (accessed on 10 August 2022).

78. Maphumulo, A.; Mahomed, O.; Vennervald, B.; Gundersen, S.G.; Taylor, M.; Kjetland, E.F. The cost of a school based mass treatment of schistosomiasis in Ugu District, KwaZulu Natal, South Africa in 2012. *PLoS ONE* **2020**, *15*, e0232867. [CrossRef] [PubMed]

79. Treatment Action Campaign. Treatment Action Campaign. Available online: https://www.tac.org.za/our-history/ (accessed on 29 January 2022).

80. World Health Organisation. Fourth WHO report on neglected tropical diseases. In *Integrating Neglected Tropical Diseases into Global Health and Development*; WHO: Geneva, Switzerland, 2017.

81. Kjetland, E.F.; Ndhlovu, P.D.; Gomo, E.; Mduluza, T.; Midzi, N.; Gwanzura, L.; Mason, P.R.; Sandvik, L.; Friis, H.; Gundersen, S.G. Association between genital schistosomiasis and HIV in rural Zimbabwean women. *Aids* **2006**, *20*, 593–600. [CrossRef]

82. Mbabazi, P.S.; Andan, O.; Fitzgerald, D.W.; Chitsulo, L.; Engels, D.; Downs, J.A. Examining the relationship between urogenital schistosomiasis and HIV infection. *PLoS Negl. Trop. Dis.* **2011**, *5*, e1396. [CrossRef]

83. Downs, J.A.; Mguta, C.; Kaatano, G.M.; Mitchell, K.B.; Bang, H.; Simplice, H.; Kalluvya, S.E.; Changalucha, J.M.; Johnson, W.D.; Fitzgerald, D.W. Urogenital schistosomiasis in women of reproductive age in Tanzania's Lake Victoria region. *Am. J. Trop. Med. Hyg.* **2011**, *84*, 364–369. [CrossRef]

84. Ndeffo Mbah, M.L.; Skrip, L.; Greenhalgh, S.; Hotez, P.; Galvani, A.P. Impact of Schistosoma mansoni on malaria transmission in Sub-Saharan Africa. *PLoS Negl. Trop. Dis.* **2014**, *8*, e3234. Available online: http://search.ebscohost.com/login.aspx?direct=true&db=mnh&AN=25329403&site=ehost-live (accessed on 19 March 2018). [CrossRef] [PubMed]

85. Bardach, E.S.; Patashnik, E.M. *A Practical Guide for Policy Analysis: The Eightfold Path to More Effective Problem Solving*, 6th ed.; CQ Press: Washington, DC, USA, 2019; pp. 1–216. Available online: https://us.sagepub.com/en-us/nam/a-practical-guide-for-policy-analysis/book255357 (accessed on 6 May 2021).

86. Bergquist, R.; Zhou, X.N.; Rollinson, D.; Reinhard-Rupp, J.; Klohe, K. Elimination of schistosomiasis: The tools required. *Infect. Dis. Poverty* **2017**, *6*, 158. [CrossRef] [PubMed]

87. Engels, D.; Wang, L.-Y.Y.; Palmer, K.L. Control of schistosomiasis in China. *Acta Trop.* **2005**, *96*, 67–68. [CrossRef] [PubMed]

88. Ross, A.G.P.; Li, Y.S.; Sleigh, A.C.; McManus, D.P. Schistosomiasis control in the People's Republic of China. *Parasitol. Today* **1997**, *13*, 152–155. [CrossRef]

89. Xu, J.; Li, S.; Zhang, L.; Bergquist, R.; Dang, H.; Wang, Q.; Lv, S.; Wang, T.; Lin, D.; Liu, J.; et al. Surveillance-based evidence: Elimination of schistosomiasis as a public health problem in the Peoples' Republic of China. *Infect. Dis. Poverty* **2020**, *9*, 39–50. [CrossRef]

90. Wang, W.; Bergquist, R.; King, C.H.; Yang, K. Elimination of schistosomiasis in china: Current status and future prospects. *PLoS Negl. Trop. Dis.* **2021**, *15*, e0009578. [CrossRef]

91. Xianyi, C.; Liying, W.; Jiming, C.; Xiaonong, Z.; Jiang, Z.; Jiagang, G.; Xiaohua, W.; Engels, D.; Minggang, C. Schistosomiasis control in China: The impact of a 10-year World Bank Loan Project (1992–2001). *Bull. World Health Organ.* **2005**, *83*, 43–48.

92. World Health Organisation. *WHO | Helminth Control in School Age Children: A Guide for Managers of Control Programmes*; World Health Organization: Geneva, Switzerland, 2016. Available online: https://www.who.int/neglected_diseases/resources/9789241548267/en/ (accessed on 24 May 2019).

93. Baan, M.; Galappaththi-Arachchige, H.N.; Gagai, S.; Aurlund, C.G.; Vennervald, B.J.; Taylor, M.; van Lieshout, L.; Kjetland, E.F. The Accuracy of Praziquantel Dose Poles for Mass Treatment of Schistosomiasis in School Girls in KwaZulu-Natal, South Africa. *PLoS Negl. Trop. Dis.* **2016**, *10*, e0004623. [CrossRef]

94. Randjelovic, A.; Frønæs, S.G.; Munsami, M.; Kvalsvig, J.D.; Zulu, S.G.; Gagai, S.; Maphumulo, A.; Sandvik, L.; Gundersen, S.G.; Kjetland, E.F.; et al. A study of hurdles in mass treatment of schistosomiasis in KwaZulu-Natal, South Africa. *S. Afri. Fam. Pract.* **2015**, *57*, 57–61. [CrossRef]

95. Corstjens, P.; de Dood, C.J.; Knopp, S.; Clements, M.N.; Ortu, G.; Umulisa, I.; Ruberanziza, E.; Wittmann, U.; Kariuki, T.; LoVerde, P.; et al. Circulating Anodic Antigen (CAA): A Highly Sensitive Diagnostic Biomarker to Detect Active Schistosoma Infections-Improvement and Use during SCORE. *Am. J. Trop. Med. Hyg.* **2020**, *103*, 50–57. [CrossRef]

96. Sousa, M.S.; van Dam, G.J.; Pinheiro, M.C.C.; de Dood, C.J.; Peralta, J.M.; Peralta, R.H.S.; Daher, E.F.; Corstjens, P.; Bezerra, F.S.M. Performance of an Ultra-Sensitive Assay Targeting the Circulating Anodic Antigen (CAA) for Detection of Schistosoma mansoni Infection in a Low Endemic Area in Brazil. *Front. Immunol.* **2019**, *10*, 682. [CrossRef] [PubMed]

97. Kvalsvig, J.D.; Schutte, C.H. The role of human water contact patterns in the transmission of schistosomiasis in an informal settlement near a major industrial area. *Ann. Trop. Med. Parasitol.* **1986**, *80*, 13–26. [CrossRef] [PubMed]

98. Sharfi, A.R.; Hassan, O. Evaluation of haematuria in Khartoum. *East Afr. Med. J.* **1994**, *71*, 29–31.

99. Ernould, J.C.; Kaman, A.; Labbo, R.; Couret, D.; Chippaux, J.P. Recent urban growth and urinary schistosomiasis in Niamey, Niger. *Trop. Med. Int. Health* **2000**, *5*, 431–437. [CrossRef] [PubMed]

100. Johnson, C.L.; Appleton, C.C. Urban schistosomiasis transmission in Pietermaritzburg, South Africa. *S. Afr. J. Epidemiol. Infect.* **2005**, *20*, 103–107. [CrossRef]

101. Agbolade, O.M.; Agu, N.C.; Adesanya, O.O.; Odejayi, A.O.; Adigun, A.A.; Adesanlu, E.B.; Ogunleye, F.G.; Sodimu, A.O.; Adeshina, S.A.; Bisiriyu, G.O.; et al. Intestinal helminthiases and schistosomiasis among school children in an urban center and some rural communities in southwest Nigeria. *Korean J. Parasitol.* **2007**, *45*, 233–238. [CrossRef]
102. Kloos, H.; Correa-Oliveira, R.; dos Reis, D.C.; Rodrigues, E.W.; Monteiro, L.A.; Gazzinelli, A. The role of population movement in the epidemiology and control of schistosomiasis in Brazil: A preliminary typology of population movement. *Memórias Do Inst. Oswaldo Cruz* **2010**, *105*, 578–586. [CrossRef]

MDPI
St. Alban-Anlage 66
4052 Basel
Switzerland
www.mdpi.com

Tropical Medicine and Infectious Disease Editorial Office
E-mail: tropicalmed@mdpi.com
www.mdpi.com/journal/tropicalmed